Multiplicity and Ontology in Deleuze and Badiou

Becky Vartabedian

Multiplicity and Ontology in Deleuze and Badiou

palgrave
macmillan

Becky Vartabedian
Regis University
Denver, CO, USA

ISBN 978-3-030-08304-5 ISBN 978-3-319-76837-3 (eBook)
https://doi.org/10.1007/978-3-319-76837-3

Softcover re-print of the Hardcover 1st edition 2018

Cover photograph by Andrew Vartabedian

Printed on acid-free paper

This Palgrave Macmillan imprint is published by Springer Nature
The registered company is Springer International Publishing AG
The registered company address is: Gewerbestrasse 11, 6330 Cham, Switzerland

For AV

Acknowledgements

This book develops themes from my doctoral dissertation, "Rethinking Multiplicity After Deleuze and Badiou," which I defended at Duquesne University in 2015. There, Dan Selcer, Jay Lampert, and Fred Evans were the first stewards of these ideas. Their attention to my work substantially improved it; I hope their influence comes through in the better parts of this text, though I am responsible for any errors and all boneheaded notions contained herein. Jim Swindal was instrumental in my arrival at Duquesne, and Joan Thompson's friendship went a long way to keep me there. I was also fortunate to have encountered a community that remains an ongoing source of support and encouragement as we've each forged ahead, philosophically and otherwise; thanks to Stephanie Adair, Taine Duncan, John Fritz, Justin Habash, Chelsea Harry, Matt Lovett, H.A. Nethery, and Kelsey Ward. "Duquesne West" was established when Boram Jeong recently arrived in Denver; much of this text was revised and improved while sharing coffee and her good company.

At Palgrave Macmillan, Phil Getz shepherded this work through the proposal and contract, and Amy Invernizzi promptly and helpfully replied to my manifold questions; I appreciate their efforts to give these ideas a home. I am also grateful to the two anonymous readers who green-lit the proposal and assessed early chapter drafts; I thank them for their attention and thoughtful comments, which have improved this text. Nathan Eckstrand offered me a chance to discuss commitments at the heart of this book on the *Blog of the American Philosophical Association* in November 2017, an opportunity I deeply appreciate.

Kent Talmage-Bowers was the first to introduce me to philosophy and its possibilities in classes at Northglenn High School almost 25 years ago, which at the time I vehemently (and somewhat legendarily) resisted. Now, my scholarly work aspires to advance Kent's insistence on understanding ideas in their context, and his practices of teaching are the bedrock of my own work with students. I also express my gratitude for the late Don McKenzie, who remains among my greatest teachers and is the inspiration for the incredulous looks I offer in response to the ambitious ideas of my own students. Karen Adkins has weathered most of my philosophical ideas, including some genuinely terrible ones about Descartes in 1999, and has always improved them; I wouldn't be where I am without her. Maria Talero thought I was fit for this style of life and I thank her for her faith. It is a pleasure to read and to think with Sarah Pessin, who brings good humor and uncommon brilliance to philosophical pursuit, and with Dan Selcer, who remains a keen interlocutor, generous collaborator, and incisive mentor.

My colleagues in the Philosophy department at Regis University, including the aforementioned Karen Adkins, Aaron Conley, Ron DiSanto, Steve Doty, Tom Duggan, Abigail Gosselin, Gregory Grobmeier, Jason Taylor, and Ted Zenzinger, each make Regis a unique place to teach and to think. Jim Seibert kindly allowed me to sit in on his 2013 "Logic and Proof" seminar and responded to left-field questions about set theory, even in the summer. Megan Patnott checked some of my work, for which I am grateful. Gabriela Carrión, Eric Fretz, Debbie Gaensbauer, Cath Kleier, Dave Law, Nate Matlock, Sally O'Laughlin, Susan Sci, Tim Trenary, Quinn and Brent Waller, and Trudi Wright are excellent colleagues and better friends.

Emily Roby skillfully clears roadblocks that would threaten my work's progress, clearing I'm unable to do on my own. The York Street crew kept the brain connected to the body and the body firmly rooted in a community for three and a half years; their good humor and camraderie supported my work. Thanks especially to Crawford Miller, Sharon Moskowitz, Andrew Sapoznik, and Erich Slouf. My life and work are rich and engaging because Kirsten Ahrendt and Alyssa Caballero Ahrendt, Pilar Allen, Justin, Tracy, and Olive Berg, Paul and Julie Cunningham, Dave DeNovellis, Molly Garrison, Matt Hrutkay and Tom Eubanks, Sarah Marvez and Donell Humphrey, and Marilyn Talmage-Bowers are each a part.

I benefit from regular care and feeding from the Case, Ketelsen, Vartabedian, and González families. Thanks to Chipper and Meggy, Patric, Kerry, Eric, Avery, Maya, and Simone; Betty and Mike, Matt, Jen, Emmett, and Maeve; and Gina, Oscar, and Pilar. I regularly care for and feed the greatest of canine companions, Lilly and Luke Vartabedian, whose requests for pets or a run outside are always well-timed. I miss my grandparents, Dariel and Nancy Case and Ed and Sue Simokat, terribly, and remain deeply grateful for the intellectual and interpersonal gifts they gave so freely.

It is a photograph of Andrew Vartabedian's that appears on this book's cover, and I'm delighted that this occasions some of our work to be presented together. He is the original reader of Deleuze and Guattari in our house, having introduced me to the notion of a rhizome on an evening walk many years ago. Andrew's curiosity propels our partnership in unexpected directions, and the intensity of his friendship is a unique pleasure. It is the greatest honor of my life to share it with him. Thank you, my dear; *you're* the best.

Contents

Abbreviations

Major Texts by Badiou

BE *Being and Event*. Translated by Oliver Feltham. London: Continuum, 2005.

CT *Briefings on Existence: A Short Treatise on Transitory Ontology*. Translated, edited, and with an introduction by Norman Madarasz. Intersections: Philosophy and Critical Theory, edited by Rodophe Gasché. Albany, NY: State University of New York Press, 2006.

DCB *Deleuze: The Clamor of Being*. Translated by Louise Burchill. Theory out of Bounds Series. Minneapolis, MN: Universit of Minnesota Press, 2000.

FP "The Fascism of the Potato." In *The Adventure of French Philosophy*, edited and translated by Bruno Bosteels, 191–201. London: Verso, 2012.

TS *Theory of the Subject*. Translated and with an Introduction by Bruno Bosteels. London: Continuum, 2009.

TW *Theoretical Writings*. Edited and Translated by Ray Brassier and Alberto Toscano. London: Continuum, 2004.

Major Text by Deleuze

DR *Difference and Repetition*. Translated by Paul Patton. New York: Columbia University Press, 1994.

Major Texts by Deleuze and Guattari

R "Rhizome: Introduction." Translated by Paul Patton. *I & C* 8 (Spring 1981): 49–71.

TP *A Thousand Plateaus, Volume 2 of Capitalism and Schizophrenia*. Translated by Brian Massumi. Minneapolis, MN: University of Minnesota Press, 1987.

WP *What is Philosophy?* Translated by Hugh Tomlinson and Graham Burchell. New York: Columbia University Press, 1994.

List of Figures

CHAPTER 1

Introduction: Lower Layers

This book investigates multiplicity in work by Gilles Deleuze (1925–1995), writing by Deleuze with his collaborator Félix Guattari (1930–1992), and in work by Alain Badiou (1937–). This investigation concerns multiplicity as an alternative to a fundamental assertion concerning the nature of being; for Deleuze writing with Guattari and for Badiou, being is neither One nor many, but *multiplicity*. In situating their respective work according to this shared commitment to multiplicity, a second commitment-in-common to multiplicity articulated with mathematical concepts and tools comes into view. Deleuze and Guattari deploy Bernhard Riemann's innovations in non-Euclidean geometry, and a particular interpretation of differential calculus. In Badiou's case, it is his well-documented use of principles at the foundation of set theory and its revisions in the late nineteenth and twentieth century, particularly Cantor's inconsistent and consistent multiples, the Zermelo–Fraenkel axiom system, and insights from Bourbaki.

This book is also a work of imagination. In the pages that follow, I invite the reader to consider a conversation that unfolded in the pages of pamphlets and texts. I reconstruct a series of exchanges between 1976 (with the publication of Deleuze and Guattari's pamphlet titled, "Rhizome—Introduction") and 1997, with Badiou's publication of *Deleuze: la clameur de l'être*, a text published in English as *Deleuze: The Clamor of Being* (2000; hereafter *Clamor*). The conversation, when it turns to questions of multiplicity and ontology, reveals objections and demands concerning the structure of multiplicity, the way certain of the

B. Vartabedian, *Multiplicity and Ontology in Deleuze and Badiou*,
https://doi.org/10.1007/978-3-319-76837-3_1

mathematical and conceptual tools are deployed to organize being qua being, and the procedures these chosen structures prescribe for handling any one in relation to this multiplicity. The prospects for approaching this as a conversation are aided considerably by my temporal distance from the original set of exchanges. I am, as it will become clear below, one in a lineage of thinkers that have taken up the Deleuze–Badiou knot; my contribution emphasizes the unique strategies each thinker takes when approaching the so-called "being question," and identifies the places where these differences give way to continuous commitments, namely the demand to arrive at and maintain the multiple using some form of subtractive procedure.

In approaching Deleuze with Guattari and Badiou at the site of multiplicity, I do so with Melville in mind; I have long been fascinated with that exchange on the deck of the *Pequod*, in which Ahab enjoins young Starbuck to "come closer … thou requirest a little lower layer." Ahab's comment is something of an existential injunction and, for the purposes of this project, a useful procedural reminder. As a reader of the Deleuze–Badiou corpus and one attentive to their exchanges, it is significant to consider how, precisely, Deleuze and Badiou each bring their reader to the site of multiplicity in their ontological projects, how they identify multiplicity with a fundamental aspect of being; and how, in Melvillian parlance, they, respectively, admonish their readers to seek this lower layer underwriting that which appears, a lower layer linked to and fundamental to its operation.

Orientations

The initiating provocation for this text is found in a different conversation, begun by the father–son team of Ricardo L. and David Nirenberg. Their 2011 article, "Badiou's Number: A Critique of Mathematics as Ontology," objects to the so-called "radical thesis" Badiou proposes in *Being and Event*, which appears in shorthand as "mathematics = ontology" (*BE* xiii). The Nirenbergs challenge the contention that mathematical ontology, in general, and especially that proposed by Badiou, produces the sorts of things it claims to; for example, they insist that Badiou's particular—and by their lights, *peculiar*—use of set theory is selective in its deployment and its consequences. The Nirenbergs claim that set theory cannot be used to justify the philosophical or political conjectures Badiou draws in *Being and Event*, and further that the identity of ontology and mathematics Badiou proposes precludes the possibility for "pathic" elements, namely human thought, to emerge from the mathematical system (Nirenberg and Nirenberg 2011, 606–612). This approach, they insist, "will entail such a drastic loss of life and experience that the result can never amount to an ontology in any humanly meaningful sense" (Nirenberg and Nirenberg 2011, 586). The Nirenbergs operate according to the view that ontology is an inquiry into being and questions related to existence as these pertain to humans; they laud the resources of phenomenology, for example, insofar as this method derives conclusions of what it is "to be" from lived experience. By emptying ontology of these resources—a traditionally human center and the lived experience that accrues to it—the Nirenbergs see Badiou's use of set theory to be so reckless as to endanger an entire tradition of thought.

The Nirenbergs' claims occasion replies from A.J. Bartlett, Justin Clemens, and Badiou himself. In a subsequent volume of *Critical Inquiry*, the discussion unfolds with accusations that one camp has fundamentally misunderstood the other. Bartlett and Clemens insist that the Nirenbergs have not read Badiou's oeuvre carefully; the Nirenbergs fail to understand that ontology as mathematics, for Badiou, *really* functions as a "figure of philosophical fiction" (Bartlett and Clemens 2012, 368). Badiou, explaining that Bartlett and Clemens respond to the Nirenbergs with "polite irony," calls the criticism raised by the Nirenbergs "stupid" (2012, 363–364). While significant points both in defense and critique of Badiou's program are raised in these pages, the interlocutors seem largely to talk past one another; the insights are, unfortunately, lost in the polemical nature of the interchange.

As I was reading this exchange, however, I felt like the conversation was missing something crucial, both as an opportunity for inquiry and a chance for defense. A closer look at the "radical thesis" Badiou proposes in *Being and Event* situates "mathematics is ontology" as the *consequent* of a preliminary claim: "Insofar as being, qua being, is nothing other than pure multiplicity, it is legitimate to say that ontology, the science of being qua being, is nothing other than mathematics itself" (*BE* xiii). Badiou spends the first five meditations of *Being and Event* arguing for the *antecedent* claim that being qua being is multiplicity. This claim is not mentioned in the Nirenberg–Bartlett–Clemens–Badiou *contretemps*; it is neither a matter of common sense nor established fact, and it led me to ask a further question following this debate: What *does* Badiou mean when he claims being qua being is pure multiplicity?

Badiou's Multiplicities

Badiou's claim that being is pure multiplicity situates his project in the debate persisting since Parmenides as to whether being is *One* or *many*. Badiou's opening salvo in Meditation One of *Being and Event* is to show this debate as having stagnated:

> For if being is one, then one must posit that what is not one, the multiple, is not. But this is unacceptable for thought, because what is presented is multiple and one cannot see how there could be an access to being outside all presentation ... On the other hand, if presentation is, then the multiple necessarily is. It follows that being is no longer reciprocal with one and thus it is no longer necessary to consider as one what presents itself, inasmuch as it is. This conclusion is equally unacceptable to thought because presentation is only this multiple inasmuch as what it presents can be counted as one; and so on. (*BE* 23)

Badiou identifies here two positions that are "unacceptable to thought." The first is the claim that being is not multiple; this is unacceptable because we experience a kind of multiplicity and diversity of things in the world. In other words, the presence of different kinds of things suggests to us that there are different ways in which a thing can be. However, and this is the second "unacceptable" claim, each of these different sorts of things presents itself to us as unified, as *one* thing. These positions are so entrenched, Badiou claims, that he must enact a *decision* that breaks the impasse and can restart ontological questioning anew. This decision consists in the claim that "the one *is not*" (*BE* 23), which means by implication that the multiple is Badiou's preferred solution to the question of being, at least as he has presented the available options.

Badiou observes a kind of dispositional affinity with the atomist programs associated with Democritus and Leucippus or the Epicureans and Lucretius, for whom 'the many' are bodies arranged in void. However, and in spite of these alignments, Badiou's pure multiple is *not* material and does *not* consist in the 'bodies' associated with these older positions. Rather, Badiou's multiple is articulated in three ways, each according to mathematical innovations: The first and second—*inconsistent* and *consistent* multiplicities—are derived from work by Georg Cantor. The third—*generic* multiplicity—comes from Paul Cohen's more contemporary work.

Georg Cantor distinguishes two types of multiplicity as a means of avoiding a common "no set of all sets" paradox in the early development of set theory, particularly with respect to the ordinal numbers (e.g., *first, second, third*). To avoid the contradiction that arises from the presence of an ordinal not counted in the set of all ordinals, Cantor posits a consistent multiple that is closed and organized according to the typical rules of the set; the inconsistent multiple, by contrast, is neither closed nor organized. It functions as an "infinity of infinities," and secures the possibility for the ordinal numbers to proceed unencumbered by closure. Badiou uses the distinction between consistent and inconsistent multiplicity to establish being qua being as an *inconsistent multiple* of precisely the type Cantor prescribes. Badiou breaks with Cantor, however, over the claim that an inconsistent multiplicity can be 'counted' by some Absolute, divine being. Instead, Badiou argues that inconsistent multiplicity is a multiplicity of *no-thing*, an open infinity that cannot be totalized. In these terms, inconsistent multiplicity constitutes "the excess presented to thought," but technically speaking inconsistent multiplicity does not make itself available to thought except through its expression in the void set, whose mark is Ø.

Consistent multiplicity for Badiou, is the mathematic template for the second aspect of the decision; if the One is not, it is still possible to claim, as Badiou does, "that *there is* Oneness" (*BE* 23). If being qua being is pure multiplicity, it is possible to derive from that pure multiple—using operations of set theory—one *as a result*. Badiou uses the axioms of Zermelo–Fraenkel (ZF) set theory to form the void set into one, which accomplishes the work of making (at least) *one* thing from an infinite, open *no-thing*, without jeopardizing the nature of inconsistent multiplicity. Badiou depends on this move to underwrite the presence of the *count-as-one* as the primary mechanism of his ontology.

Generic multiplicity, while outside the scope of this project, reflects a significant interpretation of multiplicity on which Badiou's notion of truth and the truth-procedure depend. In Meditation 33 of *Being and Event*, Badiou assigns the generic multiple as a 'being of the truth.' In other words, the generic multiple is the ontological framework that accounts for the presence of a truth, but it is not identical to the truth itself (*BE* 355). He explains that for ontology to produce a framework for generic multiplicity, it must absorb "the revolution introduced by (Paul J.) Cohen in 1963" (*BE* 355). There are several conditions that frame Badiou's work with the generic, conditions he outlines in the opening paragraphs of Meditation 33. Badiou begins by claiming, "It is impossible for mathematical ontology to dispose of a concept of truth, because any truth is post-evental, and the paradoxical multiple that is the event is prohibited from being by that ontology. The process of truth entirely escapes ontology" (*BE* 355). This claim must be read from a position where the ZF axioms apply; that is, because the matheme of the event essentially violates the rules of well-formed sets on which ZF set theory depends, a concept of truth—dependent as it is on a 'paradoxical' multiple—cannot be managed by an ontology *limited to ZF set theory*. This is intended to justify the extension to Cohen's work on the generic.

The radical thesis that animates Badiou's work—*if being qua being is pure multiplicity, then ontology is mathematics*—mobilizes a general notion of multiplicity in the three senses I've described previously. 'Pure multiplicity' is a label reserved for inconsistency, while consistent and generic multiplicities describe the mechanisms of ontology and the identification of truths, respectively. However, and as readers of Badiou's work know, this assignment is not an altogether tidy one. Although pure multiplicity is *only* inconsistent, its specter haunts ontological constructions. Consistent multiples are a means of "containing" a portion of inconsistency, expressed by its proper name Ø. However, these procedures are not always successful. An event results insofar as inconsistent multiplicity as Ø is insufficiently "contained" by procedures of consistency. Furthermore (albeit more remotely), a generic multiplicity depends on inconsistency-having-erupted for Truth and truths to proceed.

My discussion of Badiou's multiplicity focuses on inconsistent multiplicity as both foundational conceit and disruptive possibility. It is the commitment that being qua being is pure multiplicity—that antecedent of Badiou's own radical thesis—that must be tended to for the claim that mathematics is ontology to remain legible. In cases where Badiou departs from this strict presentation, he opens his ontological position to criticism; this criticism is, as I will show, potentially *more* destructive than claims about mathematics, since it is leveled at the most basic commitment underwriting the system. This weakness is exposed by several commentators, and even by Badiou himself when situating his position as an alternative to that of Heidegger, for example.[1] I offer elaboration of these criticisms and the liabilities these raise in Chapter 5.

Deleuze's Multiplicities

For Gilles Deleuze, as Manuel DeLanda (2002, 9) explains, multiplicity "is one (concept) that stands out for its longevity … (multiplicity) makes its appearance in his early books and remains one of central importance, with almost unchanged meaning and function, until his final work." Deleuze deploys multiplicity in *Bergsonism* to describe Bergson's identification of the traditional ontological problem of one and many as a 'false problem' in the history of philosophy (Deleuze 1991, 39) and devotes several pages to unpacking multiplicity as a more adequate foundation for the nature of being. In *Difference and Repetition*, multiplicity appears in the presentation of his positive position as an equivalence with the *virtual idea*. Deleuze claims, "Ideas are multiplicities; every idea is a multiplicity or a variety" (*DR* 182). The virtual idea, one half of the ontological system Deleuze develops in *Difference and Repetition* (1968/1994), is an alternative to the Kantian idea. The latter, proposed in the Schematism and appealing to the inner sense of time, fails to bridge the gap between a priori categories and the empirical intuitions the categories are said to govern. Deleuze finds this inadequate, since this process proposes the idea is a result of external conditioning—the idea only emerges after time has organized categories and intuitions.

The former, Deleuze's contribution, provides a framework for the distribution of difference *and* their relation; it is the work with the differential *dy/dx* that produces differences, and it is the framework of multiplicity that accounts for their distribution and relation. Contrary to the external conditioning of the Kantian idea, Deleuze proposes a procedure of internal genesis—that is, the idea is *expressed* by its own conditions in communication with one another; the idea is the expression, the event, of that which constitutes it. This is, conceivably, a theory of the idea 'without any gaps,' without any potential breakdowns emerging from a lapse in the inner sense of time. This is one layer of the virtual idea, or one of its philosophical valences. That the virtual idea is a *multiplicity* is perhaps trickier, since the referents include Bergson, Husserl, Riemann (multiplicity's *persona*), and Aristotle more remotely; that it appears late in *Difference and Repetition* is also a complicating factor, since the degree of philosophical complexity through which Deleuze's presentation has passed is staggering.

Whatever this virtual idea consists in, and, however, the internal genesis is to come about in relation to it, Deleuze insists that it is a multiplicity and the multiplicity must be interpreted according to a "Reimannian usage" (*DR* 182).[2] This reference points Deleuze's reader to Bernhard Riemann's analysis in the "Hypotheses" and his innovative analysis of continuous magnitudes in relation to manifolds (*Mannigfaltigkeit*). Though their presentation originates in Aristotle's *Categories*, Riemann elaborates two key features of continuous magnitudes in the 1854 text, "On the Hypotheses Which Lie at the Bases of Geometry" (hereafter "Hypotheses"): the first is the development of multiply-extended or *n*-dimensional manifolds; the second is the assertion that these are measured or compared per their own dimensionality—they are not measured according to a static or background space.

Ontologically, Deleuze follows Bergson in positing multiplicity as an alternative to *one* or *many* when it comes to assessing the nature of being. The implications of Deleuze's choice of multiplicity as an ontological framework and the nature of this multiplicity as continuous are indicated when Deleuze claims multiplicity organizes "the many as such," and that this scheme of organization "has no need whatsoever of unity in order to form a system" (*DR* 182). Continuous manifolds constitute transitions or pathways between points or subspaces in those manifolds; significantly, and as Riemann claims, the relationships constituting an extended manifold are immanently derived. By implication, these are not subject to ambient space for their determination. Furthermore, the relationships holding among points can become more complex insofar as more points or neighborhoods become available for connection.

Attention to the presence of multiplicity in *A Thousand Plateaus* (1980/1987) reveals an explicit discussion of Riemannian geometry as one among several models in "1440: The Smooth and the Striated." In this case, concerns of multiplicity and structure are re-articulated alongside several other examples, designed to illustrate the dynamic relationship between "nomad space and sedentary space" (*TP* 474), roughly spaces that allow for free circulation (smooth or nomad space) and spaces that are more rigidly organized or controlled (striated or sedentary space). In *What is Philosophy?* (1991/1994), Deleuze and Guattari weave language of *multiplicity, neighborhood*, and *dimension* into their analysis of Descartes' cogito. It is the first example of this late text; that it appears here as vocabulary key to illustrating the construction and analysis of the concept confirms its significance over the long arc of Deleuze's oeuvre. Indeed, and as I show in Chapter 4, the language of this presentation is *Riemannian* language, building a foundational concept in the history of philosophy from these convenient and long-established mathematical tools.

Riemann's work yields one additional kernel, significant for understanding Deleuze and Guattari's work with multiplicity; late in the discussion of extension, Riemann counters the $n + 1$ formula for creating dimension with a formula—marked $n - 1$—for analyzing the multiplicity. Deleuze and Guattari explicitly develop this subtractive figure in "Rhizome—Introduction" (1976/1981) and as such inaugurate a procedure of subtraction fundamental to ontological analysis in Deleuze and Guattari's work; I argue this procedure is present in Badiou's work as well.

Multiplicity and Mannigfaltigkeit

There is one additional 'sense' of multiplicity I depend on in this project, and that is its conceptual heritage as *Mannigfaltige* or *Mannigfaltigkeit*. This term is translated in Immanuel Kant's work as 'manifold,' and is identified as the predecessor of Riemann's concept of manifoldness, and subsequently informs the use of the term by Husserl and Bergson, as well as Deleuze.[3] Two versions of *Mannigfaltige* in Kant's *Critique of Pure Reason* require attention: first, *Mannigfaltige der Erscheinung*, the manifold of appearance discussed in the Transcendental Aesthetic; second, *Mannigfaltige der Vorstellungen*, the manifold of representation, discussed in the Transcendental Deduction (both A and B versions). In these cases, the manifold is associated with something external to the subject: in the Transcendental Aesthetic, the manifold of apperance is tethered to the empirical object of intuition (or the immediate object of sense-experience); in the Transcendental Deduction(s) the manifold is tethered to the transcendental object of representation, or the object = X.

The manifold of appearance characterizes the object of sense-experience. This object is undetermined, indicating that whatever it *is* is information unavailable to the subject; put another way, the "being of the manifold" is nothing other to the mind than undetermined (Kant 1998, A22/B37). As an undetermined field of data, the manifold of appearance is structured from without. The mind, in its positing the structure of space, determines objects according to other objects, using tools of "form, magnitude, and relation" (Kant 1998, A22/B37). Perhaps peculiarly, Kant (1998, A28/B44) argues that space has both "empirical reality" insofar as it structures "all possible outer experience" and "transcendental ideality," insofar as it "is nothing as soon as we leave out the condition of the possibility of all experience." As an intuited condition of the mind, space organizes the manifold of outer appearances according to their simultaneity, or appearing together; the a priori intuition of time, as it is presented in the Transcendental Aesthetic, organizes appearances according to succession. These structures shape experience (its empirical reality), but are not attributable to objects themselves—only to minds that engage the manifold (its transcendental ideality). The manifold thus has no power to organize itself; some structuring intellect is required to bring organization and determination to the undetermined. The work of the Transcendental Aesthetic reveals that objects aren't actually experienced per se, but rather the manifold as an undetermined field of data is structured (using outer and inner senses of space and time, respectively) by the mind according to determinate relations of succession and simultaneity.

The second version of the manifold deals with the re-presentation of the manifold in the understanding. In the A Deduction, the object is relegated to 'placeholder,' or object = *X*, while the understanding applies the categories to the object = *X*.[4] The categories give the conditions by which objects can be organized in addition to the resources of space and time, and lend a semblance of objectivity to experience, thus providing the conditions by which judgments about objects are possible. In other words, the faculty of sensibility—that initial, sensory encounter with an object—simply identifies that "something" is out there; the categories describe that "something out there" according to more specific parameters.

Before arriving at the object = *X*, Kant (1998, A104) begins by describing the "objects of representations," which "must not be regarded in themselves," but rather regarded only according to the way they are represented. While objects cannot be known or their essence understood by cognition, the object, in general, functions as the requisite source of corresponding representations. On account of this dual edge—both the necessity of the object and the limitation of cognition to apprehend it in itself—Kant (1998, A104) says, "It is easy to see that this object must be thought of only as something in general = *X*, since outside of our cognition we have nothing that we could set over against this cognition as corresponding to it." The object = *X*, then, functions as a placeholder such that my representation of any object's appearance might be continuous. Exterior to the work of cognition, there are no determinations of objects whatever. This limitation appears as early as the discussion of appearance in the Transcendental Aesthetic, but Kant insists that the object = *X* is necessary for underwriting the continuity of representation (1998, A105).

This object = *X*, Kant (1998, A105) explains, is only a correspondent to our representations and, "because it should be something distinct from all our representations, is nothing for us." Like the content of the manifold of appearance, the object is necessary for the continuity of representation, but remains inaccessible to the faculty of understanding. However, the object as *that which is known* necessitates a unity, and this unity "can be nothing other than the formal unity of the consciousness in the synthesis of the manifold of the representations" (Kant 1998, A105). A theme recurs here—the object is present as a necessary condition for the unity of representation, but the object gains no further definition qua object in this analysis. The object of intuition persists as a condition for continuous cognition, but beyond that, it remains undetermined. Additionally, Kant (1998, A109) explains that the work of cognition can bring the understanding to an objective reality in which "all of our empirical concepts can provide relation to an object." However, this *relation* marks the boundary of cognition, since the object is the object qua transcendental object = *X* (Kant 1998, A110). The aim of the A Deduction is, in part, to explain the conditions by which cognition is possible at all.

The B Deduction opens with an investigation into the "possibility of a combination in general" (B129), accomplished by the 'interior' functions organizing the manifold rather than positing the transcendental object = *X* as an exterior source and limit of cognition. However, Kant's claims in the B Deduction are consonant with those made in the A Deduction insofar as the object and its qualities mark the boundary of cognitive analysis. The understanding organizes and determines that which is represented to it; it cannot go beyond this data toward any conclusion about the re-presented manifold in itself. The presentation of manifold in the A and B Deductions above reinforces the claim that the object of cognition is unknown in itself, and in spite of its elevation to transcendental object or an object, in general, its status remains undetermined except in connection with the actions of a cognizing subject.

Though I've focused here on the object = *X*, this is embedded in the broader project of explaining the scaffolding of cognition down to its 'root' in the transcendental unity of apperception, the general mode of reflective subjectivity required to cognize at all. This unity allows Kant to posit a 'transcendental subject,' or the unity of consciousness that makes all cognition possible. In his monograph on Kant, Deleuze (1984, 16) presents this in a kind of formula, saying "I think myself and in thinking myself, I think the object in general to which I relate a represented diversity." Just as the Transcendental Aesthetic explained the encounter of sensibility with an object, the A Deduction explains the necessity of a transcendental object (the object = *X*) to hold the attention of the cognizing subject.[5]

The manifold has a clear epistemological function in Kant's first *Critique*; it is that which the mind can determine according to inner and outer senses (time and space, respectively) and the categories (which, by the argument in the B Deduction, are shown as judgments of the understanding), as well as hold as the condition for the possibility of any representation whatsoever (i.e., the transcendental object = *X* argument in the A Deduction). The ontological sense of the manifold, however, is not only undetermined but also *indeterminate*. The object itself forms a boundary that cognition simply cannot cross, and whatever is on the other side of this boundary is off limits for *ontological* determination. Though Deleuze explicitly engages Kant's notion of manifold and his deployment of the transcendental object = *X*, I am especially interested in considering Kant's framework of boundary and structure as these pertain to the structure and specific commitments presented in Badiou's ontology. I address this relationship in Chapter 5.

Method

In reading work by Badiou and Deleuze over shared conceptual commitments like *event, subject*, or *multiplicity*, there seem to be four lines of approach. A first approach is to focus or specialize in the work of one thinker, to master the long arc of relation holding between key concepts. The tradition of scholarship of this sort on Deleuze is deep; reflection on Badiou's work continues to grow. For example, Peter Hallward (2003) offers a comprehensive accounting of Badiou's position, and as one of the first studies of Badiou's work of length in English, continues to serve as a key introduction to Badiou's work. Readers find, for example, Oliver Feltham (2008) and Ed Pluth (2010) each offering sustained engagement with key concepts over the long arc of Badiou's productive career.

A second requires recognizing charges of inadequacy levied by one thinker against the other, an approach that has mainly consisted in thoughtful defenses of Deleuze's oeuvre against Badiou's apparent misreading and misunderstanding of key commitments in Deleuze's work. English-language readers see, for example, Daniel W. Smith's "Mathematics and the Theory of Multiplicities: Badiou and Deleuze Revisited" ([2003] 2012) as an entry defending and clarifying Deleuzian commitments in light of their critique in Badiou's monograph on Deleuze, published in English in 2000 as *Deleuze: The Clamor of Being* (hereafter *Clamor*). More recently, Jon Roffe (2012) and Clayton Crockett (2013) each offer book-length treatments that take Badiou's interpretation in *Clamor* to task. A third approach situates these thinkers generationally or as "backstops" for reflection; as John Mullarkey (2003) and Ian James (2012) recognize, Deleuze's work is of a generation older than Badiou and his contemporaries. In this framework, Deleuze's work is presented as a kind of heritage position with which thinkers in the succeeding generation must contend. Sam Gillespie (2008) begins with a reflection on Badiou's and Deleuze's commitments-in-common to multiplicity and uses this common ground to discuss Badiou's unique ontological direction.

In this text, I follow a fourth approach: I stage a conversation between Badiou and Deleuze. The topic of this conversation is multiplicity, and particularly its presence in and object of ontological inquiry. In this trajectory, my work follows the model presented by Simon Duffy (2013, 137), who affirms an approach that "(reads) their respective interpretations of mathematics, and the role that they each assign to mathematics in the development of their respective philosophical projects, together, alongside of one another." Duffy insists that this approach identifies "points of convergence," in addition to sharpening the sites of divergence that obtain between the two programs.

The evidence for this conversation, which I develop with some care in the next chapter, is found in a series of exchanges in the period between 1976 and 1997, the period inaugurated with Deleuze and Guattari's "Rhizome—Introduction," and ended with the French publication of Badiou's monograph on Deleuze. In this period, I identify several exchanges that take concerns of multiplicity and its relationship to any emerging or static 'one' as their center; these exchanges are largely indirect, appearing 'after the fact' or in response to the work of the other. To see this as a conversation, however, it is insufficient to attend to the content of the exchanges alone; rather, the deep commitments with which the interlocutors arrive to the exchange must be set out. In this case, an appreciation of Badiou's and Deleuze's respective accounts of multiplicity and the mathematical paradigms on which these accounts depend serve to deepen the conversation in the period I've identified above, and particularly as these pertain to the development of Badiou's ontology, the structure of static genesis or points associated with the place of the 'virtual' in Deleuze's solo work, and the structure of the rhizome and concept in Deleuze and Guattari's collaboration. To this end, I engage work by Riemann, Cantor, Zermelo, and Bourbaki in a way that appreciates their technical specificity, and so I discuss the original sources and key concepts at stake, but I have endeavored to make their presentation clear and straightforward for the non-specialist reader.[6]

The conversational strategy resists a tendency, apparent in Badiou's work with other philosophers, described by Bruno Bosteels (2012, ix) as "constitutive polemical knots." These are argumentative strategies that, per Bosteels (2012, ix–x),

> ... give Badiou's philosophy its distinctive orientation, tonality, and feel. Indeed, one of this thinker's greatest virtues – which to others might seem a defect, especially in his writings on other philosophers – lies in giving thought a decisive orientation by leading readers to the point where they must take a stand in one way or another.

Badiou's relationship to Deleuze represents one of the major polemical knots organizing his work. This relationship is confrontational along the lines Bosteels describes. In the Introduction to *Clamor*, Badiou continuously casts his program against Deleuze's, identifying "an immanent conception of the multiple" (*DCB* 4) as the central pole around which their respective work orbits. Though my work proceeds by taking this opposition seriously, I reject the requirement that the opposition must be resolved in favor of one thinker over the other. The nature and features of the multiple are precisely the elements under scrutiny, and the demand Badiou makes of the reader requires it be comprehensively established that the nature of being qua being is as Badiou characterizes it, thus pre-empting the possibility for conversation altogether.

Given the rich resources emerging in the wake of *Clamor*, the student of the Badiou–Deleuze disjunction may take up a slower, focused strategy that reads multiplicity and its link to being qua being deeply and according to the unique commitments attending each program. This allows, I propose, not only a better understanding of one program vis-à-vis the other, but also a broader frame of understanding. This broader frame appreciates political, conceptual, and procedural commitments, and thus imagines a focused conversation in which both interlocutors are able to "speak their piece," an approach that students now (at least) two generations of scholarship removed from the original interchanges are able to take up; indeed, it is only because of the breadth of work developed in the three approaches I describe above that the tack I take is possible.

Though readers of Badiou and Deleuze together are right to see their approaches as largely incommensurate, owing to the diverging interpretations of multiplicity developed in their respective accounts and the differing mathematical paradigms in action, my contribution identifies places where these approaches give way to continuous *procedural* commitments—that is, not only do Badiou and Deleuze share a commitment-in-common to a form of multiplicity in their ontological accounts, but they both prescribe a requirement to maintain the presence of this multiple using some form of subtractive procedure. Additionally, seeing these thinkers in conversation clarifies the objections—which remain remarkably similar from the beginning of the conversation to its end—one thinker maintains against the other. As I point out in this work's Conclusion, the consistency found in these critiques is a way of marking the hazards attending an ontology developed without a subject at its center.

Plan of the Text

There are three claims that structure my work in this text. First, I insist that an analysis of multiplicity in work by Alain Badiou, Gilles Deleuze, and in Deleuze's collaboration with Félix Guattari, clarifies key moves or signposts of their respective theories. Second, articulating each of the programs on their own terms allows readers to untangle the polemic that often accompanies (and sometimes obscures) the analysis of multiplicity. Third and finally, and as this work's Conclusion demonstrates, attention to multiplicity gives way to fundamental questions about the function of *decision* and *dissolution* in ontological discourse. That is, accompanying the presentation of multiplicity is a schema for determining what 'counts' for being qua being in the first place. This is the 'lowest layer' this investigation proposes to uncover.

In service of this threefold strategy, I begin with "Engagements." Chapter 2 tracks Badiou's and Deleuze and Guattari's presentations of multiplicity in the period 1976–1997. It is Deleuze and Guattari's "Rhizome," the pamphlet published in 1976 and destined to become the introduction to *A Thousand Plateaus*, that I identify as inaugurating this conversation. I focus on two accomplishments in this pamphlet: first, the critique Deleuze and Guattari raise against 'traditional' forms of materialism (in general) and Maoism (in particular); second, their peculiar articulation of multiplicity as the third principle and structure of the alternative they propose. In these two accomplishments, readers find the terms on which Badiou will object to Deleuze and Guattari (and vice versa): *procedure* and *structure*. Procedure concerns the relationship of unity to multiplicity, in terms of production and priority; in other words, critiques of procedure are directed at the way any one emerges from or is cut from a multiple, and the status that one has with respect to the multiple from which it is drawn. Structure describes the nature of the multiplicity; it is in articulation of structure that the mathematical concepts each thinker depends on become clear, and while I mention these in Chapter 2, I reserve full investigation of these mathematical underpinnings for Chapter 3.

The two prongs of procedure and structure are again at issue in Badiou's critique of Deleuze and Guattari's *Rhizome*. Writing pseudonymously as "Georges Peyrol," Badiou's 1977 "The Fascism of the Potato" responds to the critiques Deleuze and Guattari direct at Maoist dialectical procedure and the structure of the multiple. Badiou/Peyrol returns fire, condemning the rhizomatic structure Deleuze and Guattari propound for its unsatisfactory political consequences. Reflection on the failure of May '68 leads Badiou/Peyrol to argue that the absence of any unifying force cut from the masses (e.g., class) will result in "the flattest of conservatisms" (*FP* 196), the opposite of any vigorous materialist revolt.

I then fast-forward to 1991, in which Deleuze and Guattari make direct address of Badiou's ontological proposal in "Example 12" of their *What is Philosophy?* Here, Deleuze and Guattari assess Badiou's work in their appraisal of logic, or a system of thought that *appears* to be philosophical but is instead tethered to science, its fixed frames of reference, and propositional claims. Logic and its prospects—those elements from which a logical system is built—fail to appreciate the creative and transient nature of the concepts on which philosophy relies. Once again, the critique Deleuze and Guattari raise is couched in terms of structure (in this case the 'place' of logic's origination as unspecified multiplicity) and procedure (or the construction of sets from this unspecified multiplicity). I conclude by assessing Badiou's critique of Deleuze in *Clamor*. This text argues that for all of his conceptual innovation and dedication, Deleuze is ultimately a philosopher of the One. I survey the critique in Badiou's text, the critique Badiou's text occasions, and note the ways these highlight the structure-procedure elements that characterize the conversation up to *Clamor*.

The work in Chapter 2 sets the limits for texts under investigation in this study. *Clamor* may be identified as the 'last word' in the sustained conversation between the Badiouian and Deleuzian camps, a conversation closed by Deleuze's death in 1995. However, the parameters for my inquiry are effectively limited by Deleuze and Guattari's remarks on Badiou's work in *What is Philosophy*? Though Badiou's commitments to axiomatic procedures are evident in work as early as 1968–1969, it is reasonable to conclude that Deleuze and Guattari respond in "Example 12" to the mature presentation of Badiou's system in *Being and Event*, published as *L'être et l'événement* in 1988.

To be sure, in the 20 years following the initial publication of *Deleuze: la clameur de l'être*, Deleuze continues to offer Badiou a touchstone for reflection; readers of Badiou's *Logics of Worlds* (2009a) find reflections on Deleuze's conception of the event, and in a paean to Deleuze in *Pocket Pantheon* (2009b), a meditation on their respective rejections of finitude. Badiou's elaboration of his system in *Logics of Worlds* and the forthcoming *Immanence of Truths* must necessarily engage conceptions of multiplicity built from the earlier text. However, I do not address these more contemporary reflections but with two exceptions: first, "The Question of Being Today," an essay opening Badiou's 1998 *Court Traité d'ontologie transitoire*, where he elaborates the nature of inconsistent multiplicity; and second, the address titled "One, Multiple, Multiplicities," appearing originally in *Multitudes* (2000) as a response to critics of *Clamor*. Badiou directly engages the Riemannian underpinnings of Deleuze's system of multiplicity and in some ways, is a fulfillment of the promise of *Clamor*. I discuss this text, effectively an epilogue to the main engagements of 1977–1997, in Chapter 5.

As the first of two chapters devoted to an explication of multiplicity and its subsequent clarification of key elements of Deleuze's and Badiou's oeuvres Chapter 3, titled "Structures," attends to multiplicity-as-structure by focusing on the respective mathematical commitments on which these thinkers depend. I begin by describing Bernhard Riemann's "Hypotheses" and two innovations therein: his analysis of continuous and discrete manifoldness (itself an extension of the original Aristotelian distinction), and the notion of multiply-extended manifoldness. I discuss a third result of Riemann's investigation, an alternative comprehension of or disposition toward physical space. I then read these back into Deleuze's theory of the virtual idea as it is explicitly defined in *Difference and Repetition*'s chapter 4 and in the analysis of space in Deleuze and Guattari's Plateau "1440: The Smooth and the Striated." Attention to Riemann and his contributions assists Deleuze's reader in confirming the consistency of his commitment to multiplicity as an ontological structure.

I then turn to Badiou's theory of multiplicity, with a special focus on the distinction of inconsistent and consistent multiplicities. Badiou establishes this early in *Being and Event* and it remains the foundational distinction on which subsequent invocations of transfinite numbers (cf. Meditations 12–14 of *BE*), situation, and generic multiplicity depend. I explain the Cantorian origination of this distinction, focusing on his solution to the Burali-Forti Paradox. I then explain how this initial distinction is decisive for understanding the structure of Badiou's multiplicity and foregrounds his continuing commitments to a form of materialism.

With a thorough accounting of multiplicity-as-structure accomplished, Chapter 4, "Procedures," assesses the techniques by which relationships of one and multiple are determined in these systems. I begin by discussing the presentation in *Being and Event* of the void set and its status as existential axiom, key in building any situation from the multiple, or in making the link of consistency to inconsistency. I then attend to the analysis of scission in Badiou's 1975 lectures, collected in *Theory of the Subject*; this offers a preliminary sketch of procedures for determining an existent from a world. I address the differences between the project relating one to multiple in *Being and Event* and the earlier one proposed in *Theory of the Subject* with the aim of discussing themes in-common between Badiou's explicitly ontological and political works. I then turn to the subtractive procedure Deleuze and Guattari propose in "Rhizome," explaining the content of this procedure and its prescriptions for understanding the way any unity is positioned in Deleuze and Guattari's system. Subtraction is put to effective use in understanding the way a concept is constructed, and so I discuss "Example 1" to apply the insights of subtractive procedure to the discussion of Descartes' cogito in *What is Philosophy?*

Attention to their respective theories of multiplicity and mathematical underpinnings clears the way for a precise return to the objections concerning structure and procedure, a reconstruction that is the subject of this work's Chapter 5. In "Re-engagements," I begin by discussing two forms of subtraction in Badiou's work: first, the 'special' presentation in the Gamma Diagram; second, its 'general' form in a retrospective procedure revealing the presence of inconsistency in any situation. I then demonstrate certain challenges that attend Badiou's later ontological articulations, particularly in "Kant's Subtractive Ontology" and in *Briefings on Existence* connected to his analysis of the fundamental ontological choice. The presence of a general subtractive analysis, especially when considered in combination with scission, allows me to identify two, perhaps unexpected, continuities emerging from the analysis of subtraction in Deleuze and Guattari's work. In both programs, subtraction establishes a *means of access* to the structure of multiplicity; subtraction also delivers a *mode of maintenance* designed to 'keep' multiplicity as the ground or foundation of ontological investigation. I will show that both Deleuze (with Guattari) and Badiou depend on subtraction as a mechanism for managing the relationship of singular instances (in the former) and count-structures (in the latter) to the multiplicity from which they are derived.

I then turn to the explicit critiques raised in Chapter 2's history, and I demonstrate that the resources a thoroughgoing analysis of multiplicity as both structure and procedure offer for understanding and evaluating these critiques. I reorient the 'direction' of multiplicity in a counter to Deleuze and Guattari's critique of arborescence, and subsequently discuss the claims to rigidity that Deleuze and Guattari attribute to Badiou's project. I then assess the critique of 'cases' Badiou raises against Deleuze in *Clamor*, in which Badiou claims that all unique instances are fodder for Deleuze's theory and are thus emptied of their significance. Aided by insights from Simon Duffy's analysis of Deleuze's modes of using mathematics, I respond to this critique and demonstrate the divergence required by Badiou's adherence to the "radical thesis."

After a review of the conversation I staged in the main body of the text, I conclude by considering an engagement that does not appear in the preceding: Badiou's 1988 review of Deleuze's *The Fold*. The brief and vigorous critique of multiplicity Badiou offers in this text opens on to the most fundamental of all commitments for an ontology without subjectivity: a *decision*, in which the program must re-form elements of what is already available, or a *dissolution*, in which the program reinvents the field and relations of elements entirely.

Notes

1. See, in addition to the aforementioned work by the Nirenbergs, Adrian Johnston (2008) and Geoffrey Pfeifer (2015).
2. Riemann's name is misspelled when it appears in Patton's translation on p. 182. It is subsequently corrected in Patton's text.
3. Daniel W. Smith (2012, 123) reads the lineage of this concept backward from Deleuze's usage of multiplicity, noting it "was first formulated mathematically by Bernard (*sic*) Riemann, in his non-Euclidean Geometry, who in turn linked it to Kant's concept of the manifold. Both Husserl and Bergson adopted Riemann's concept for their own philosophical purposes, and Deleuze first wrote about the concept with regard to Bergson's distinction between two types of multiplicity (continuous and discrete), which Deleuze developed in his own manner and considered it one of the fundamental problems of contemporary thought." The relations Smith identifies here accounts for Deleuze's assertion in *Difference and Repetition* that his own concept of multiplicity is continuous with that of Bergson, Husserl, and Riemann. M.D. Maia (2011, 9) offers an account of the relation of Riemann's work to Kant's noting that Riemann's use of *Mannigfaltige* has "a slightly different meaning to define his metric geometry."
4. At A111 Kant argues, "the categories that have just been adduced are nothing other than the conditions of thinking in a possible experience, just as space and time contain the conditions of intuition for the very same thing They are therefore also fundamental concepts for thinking objects in general for the appearances, and they therefore have a priori objective validity, which was just what we really wanted to know." The analysis of the transcendental object = *X* is a vehicle by which Kant can arrive at the demonstration of the categories, the second task of the deduction after the argument for the transcendental unity of apperception.
5. In *Kant's Critical Philosophy* Deleuze explains that synthesis necessitates the presence of an object 'in general' in order to accomplish its task. He says, "the manifold would never be referred to an object if we did not have at our disposal objectivity as a form in general ('object in general', 'object = x'). Where does this form come from? The *object in general* is the correlate of the 'I think' or the unity of consciousness; it is the expression of the cogito, its formal objectivation" (15).

6. Baki (2015) offers a comprehensive and technical accounting of the mathematics at stake in Badiou's *Being and Event*. Duffy (2013), as I have explained above, elaborates a series of mathematical and philosophical engagements in Deleuze's work.

References

Badiou, Alain. 2000. *Deleuze: The Clamor of Being*. Translated by Louise Burchill. Theory Out of Bounds Series. Minneapolis, MN: University of Minnesota Press. Originally *Deleuze: la clameur de l'être*. Paris: Hachette, 1997. (Cited in text as *DCB*)

———. [2000] 2004. "One, Multiple, Multiplicities." In *Theoretical Writings*, edited and translated by Ray Brassier and Alberto Toscano, 67–80. London: Continuum. Originally "Un, multiple, multiplicite(s)." In *Multitudes* 1 (2000): 195–211.

———. 2005. *Being and Event*. Translated by Oliver Feltham. London: Continuum. Originally *L'être et l'événement*. Paris: Éditions du Seuil, 1988. (Cited in text as *BE*)

———. 2009a. *Logics of Worlds: Being and Event II*. Translated by Alberto Toscano. London: Bloomsbury. Originally *Logiques des mondes*. Paris: Éditions du Seuil, 2006.

———. 2009b. *Pocket Pantheon: Figures of Postwar Philosophy*. Translated by David Macey. London: Verso. Originally *Petit Panthéon Portatif*. Paris: Éditions la Fabrique, 2008.

———. 2009c. *Theory of the Subject*. Translated and with an Introduction by Bruno Bosteels. London: Continuum. Originally *Théorie du sujet*. Paris: Éditions du Seuil, 1982. (Cited in text as *TS*)

———. 2012. "The Fascism of the Potato." In *The Adventure of French Philosophy*, edited and translated by Bruno Bosteels, 191–201. London: Verso. Originally Georges Peyrol. "Le fascisme de la pomme de terre." In *La situation actuelle sur le front de la philosophie*, 42–52. Paris: François Maspero, 1977. (Cited in text as *FP*)

Badiou, Alain, A.J. Bartlett, and Justin Clemens. 2012. "I: To Preface the Response to the 'Criticisms' of Ricardo Nirenberg and David Nirenberg." *Critical Inquiry* 38 (2): 362–364. https://doi.org/10.1086/662746. Accessed 2 August 2012.

Baki, Burhanuddin. 2015. *Badiou's Being and Event and the Mathematics of Set Theory*. London: Continuum.

Bartlett, A.J., and Justin Clemens. 2012. "Critical Response II: Neither/Nor." *Critical Inquiry* 38 (2): 365–380. https://doi.org/10.1086/662747. Accessed 1 September 2013.

Bosteels, Bruno. 2012. "Translator's Introduction." In Alain Badiou, *The Adventure of French Philosophy*, edited and translated by Bruno Bosteels, vii–lxiii. London: Verso.

Crockett, Clayton. 2013. *Deleuze Beyond Badiou: Ontology, Multiplicity, and Event*. New York: Columbia University Press.

Delanda, Manuel. 2002. *Intensive Science and Virtual Philosophy*. London: Continuum.

Deleuze, Gilles. 1984. *Kant's Critical Philosophy: The Doctrine of the Faculties*. Translated by Hugh Tomlinson and Barbara Habberjam. Minneapolis, MN: University of Minnesota Press. Originally *La Philosophie Critique de Kant*. Paris: Presses Universitaires de France, 1963.

———. 1991. *Bergsonism*. Translated by Hugh Tomlinson and Barbara Habberjam. New York: Zone Books. Originally *Le Bergsonisme*. Paris: Presses Universitaires de France, 1966.

———. 1994. *Difference and Repetition*. Translated by Paul Patton. New York: Columbia University Press. Originally *Différence et repetition*. Paris: Presses Universitaires de France, 1968. (Cited in text as *DR*)

Deleuze, Gilles, and Félix Guattari. 1981. *Rhizome: An Introduction*. Translated by Paul Patton. *I & C* 8 (Spring): 49–71. Originally *Rhizome: Introduction*. Paris: Les Éditions de Minuit, 1976. (Cited in text as *R*)

———. 1987. *A Thousand Plateaus*. Translated by Brian Massumi. Minneapolis, MN: University of Minnesota Press. Originally *Mille Plateaux*. Paris: Les Éditions de Minuit, 1980. (Cited in text as *TP*)

———. 1994. *What Is Philosophy?* Translated by Hugh Tomlinson and Graham Burchell. New York: Columbia University Press. Originally *Qu'est-ce que la philosophie?* Paris: Les Éditions de Minuit, 1991. (Cited in text as *WP*)

Duffy, Simon. 2013. *Deleuze and the History of Mathematics: In Defense of the "New"*. Bloomsbury Studies in Continental Philosophy Series. London: Bloomsbury.

Feltham, Oliver. 2008. *Alain Badiou: Live Theory*. London: Continuum.

Gillespie, Sam. 2008. *The Mathematics of Novelty: Badiou's Minimalist Metaphysics*. Anamnesis. Melbourne: re.press.

Hallward, Peter. 2003. *Badiou: A Subject to Truth*. Minneapolis, MN: University of Minnesota Press.

James, Ian. 2012. *The New French Philosophy*. London: Polity.

Johnston, Adrian. 2008. "Phantom of Consistency: Alain Badiou and Kantian Transcendental Idealism." *Continental Philosophy Review* 41: 345–366. https://doi.org/10.1007/s11007-008-9086-5. Accessed 10 March 2015.

Kant, Immanuel. 1998. *Critique of Pure Reason*. Translated by Paul Guyer and Allen W. Wood. Cambridge, UK: Cambridge University Press. German references are to *Kritik der Reinen Vernunft*. Leipzig, 1919.

Maia, M.D. 2011. "The Physical Manifold." In *Geometry of the Fundamental Interactions*, 9–24. https://doi.org/10.1007/978-1-4419-8273-5_2. Accessed 6 January 2014.

Mullarkey, John. 2003. *Post-continental Philosophy: An Outline*. London: Continuum.

Nirenberg, Ricardo L., and David Nirenberg. 2011. "Badiou's Number: A Critique of Mathematics as Ontology." *Critical Inquiry* 37 (4): 583–614. https://doi.org/10.1086/660983. Accessed 31 August 2011.

Pfeifer, Geoffrey. 2015. *The New Materialism: Althusser, Badiou and Žižek*. London: Routledge.

Pluth, Ed. 2010. *Badiou: A Philosophy of the New*. Key Contemporary Thinkers series. London: Polity.

Roffe, Jon. 2012. *Badiou's Deleuze*. Montréal: McGill-Queen's University Press.

Smith, Daniel W. [2003] 2012. "Mathematics and the Theory of Multiplicities: Badiou and Deleuze Revisited." In *Essays on Deleuze*, 287–311. Edinburgh: Edinburgh University Press.

———. 2012. "Analytics: On the Becoming of Concepts." In *Essays on Deleuze*, 122–145. Edinburgh: Edinburgh University Press.

CHAPTER 2

Engagements, 1976–1997: History of a Misunderstanding

This chapter develops a conversation concerning the nature of multiplicity and its relationship to one, drawing from material written by Deleuze with Guattari and situates Badiou. This conversation begins in the aftermath of the May '68 uprising, in which thinkers on the left are working to recover and restate forms of materialism and leftism. It is in this intellectual context that Deleuze and Guattari pen "Rhizome-Introduction" ([1976] 1981), which offers an alternative vision of materialism, one emphasizing self-organizing and self-regulating systems. Their text occasions a reply, titled "The Fascism of the Potato," written by Georges Peyrol (a pseudonymous Alain Badiou) in 1977. Badiou/Peyrol's text critiques Deleuze and Guattari's work with attention to *structure* (the way a multiplicity is composed) and *procedure* (the relation of one to multiple prescribed by this structure). I begin by discussing Deleuze and Guattari's presentation and Peyrol's subsequent critique. I use the terms *structure* and *procedure* discerned in Peyrol's work to organize the subsequent stages of this conversation and my larger investigation.

I suggest that using *structure* and *procedure* as an analytical framework illuminates readings of Badiou's work by Deleuze and Guattari in *What is Philosophy?* ([1991] 1994) where the authors address the prospects of Badiou's ontology in *Being and Event* and in terms of the multiple he proposes. This framework also offers another way into Badiou's *Deleuze: the Clamor of Being* ([1997] 2000), a text rich in readings—and misreadings—of Deleuze's philosophical project. In this late entry, which effectively 'seals' the conversation between Badiou and Deleuze, I show

B. Vartabedian, *Multiplicity and Ontology in Deleuze and Badiou*,
https://doi.org/10.1007/978-3-319-76837-3_2

that Badiou's primary critique is structural and thus rehearses the early challenges Peyrol raised against Deleuze and Guattari in "Rhizome." This insight prefaces the assessment of structure in Deleuze's and Badiou's respective work in Chapter 3. The reader will note that this chapter is not concerned with whether one thinker "got it right" when engaging the work of the other. Rather, in the set of engagements I unfold here, I am interested to identify the main criticisms one thinker brings against the other(s), and the standard modes of reply when critically engaged.

1976–1977: Root, Rhizome, Potato

The conversation concerning multiplicity emerges in an already combative climate. Since leftism (broadly), Communism and Maoism (more specifically) each failed to mobilize the revolutionary spirit of May '68 toward transformation, proponents of these positions were required to rethink their commitments. Some, like the so-called "New Philosophers," turned vehemently against their past efforts, disavowing their participation in revolutionary activities, and claiming that the theoretical commitments underpinning the movement were Stalinist in their trajectory.[1] For their part, Badiou and Deleuze (writing primarily in this period with Guattari) each maintain a fidelity to materialism, albeit in different expressions.

In this section, I discuss Badiou's and Deleuze and Guattari's *après-Mai* revisions of their materialist commitments, beginning with Deleuze and Guattari's "Rhizome—Introduction" ([1976/1981], hereafter "Rhizome"). In this text, their commitments to materialism are bound up with an account of multiplicity. I then discuss two of Badiou's 1977 replies: first, and briefly, "Flux and the Party," a critique of the program Deleuze and Guattari propose in *Anti-Oedipus*; second, and more substantively, "The Fascism of the Potato," which Badiou penned as 'Georges Peyrol.' In this latter text, Badiou/Peyrol takes aim at Deleuze and Guattari's presentation of *structure*, the composition of multiplicity itself, and *procedure*, or the relationship between one and multiple they propose.

1976: Rhizome

Deleuze and Guattari's "Rhizome" is the first salvo in what will become a two-decades-long exchange with Badiou. Deleuze and Guattari critique a relation of matter to thought at the heart of traditional materialism expressed in the slogan "the One that becomes two." Before presenting multiplicity as an alternative means of characterizing the relation of thought and matter, Deleuze and Guattari begin with the image of thought as a *book* and offer two accounts of its relation to its source material, or *nature*: first, as a unified form of imitation; second, as a systematic work of creation.

Deleuze and Guattari explain that the classical book "imitates the world, as art does nature: by procedures which are peculiar to it, and which carry out what nature cannot or can no longer do" (*R* 51). As an imitation, or instrument of reflection, the book represents material conditions in thought. However, and as Deleuze and Guattari point out, claims to imitation are complicated by the fact that nature fails to reproduce itself with the regularity and precision prescribed by the image of the *book-as-imitation*. The book also articulates a materialist law, *the One which becomes two*. The "One" to which this law refers is the fundamental claim of materialism that what *is* is matter; the "two" is the reflection of matter in thought. At stake, however, and at the heart of Deleuze and Guattari's criticism, is that this theory is dependent on some arbiter of accuracy; put another way, traditional materialism is beholden to an external measure—a 'third man' of sorts—adequate to evaluate the precision of the reproduction.

This commitment to accuracy in reflection depends on the production of a dichotomy between thought and nature (i.e., "thought is not matter"). Deleuze and Guattari thus claim this position " … needs the presupposition of a strong principal unity to arrive at two by following an ideal method" (*R* 51). The principal unity Deleuze and Guattari assign to this position is the fundamental assumption that whatever is *is* matter. The dichotomous claim, *thought is not matter* thus requires correlate support in the claims *thought is adequate because it is sufficiently like matter* and *matter exists before thought*. The priority of matter is the claim required for initiating a sequence of thought-production.

In this context, the story Deleuze and Guattari tell about the book situates it as a second-order reflection; the One of matter and the two of the thought-matter reflection together give way to the 'three' of the book as the record of this thought-matter reflection. Readers of Deleuze and Guattari may identify a procedure for *counting* here, a scheme of reproducing the original and initial One. Take, for example, Jacques Lacan's 1966 Baltimore Lecture, "On Structure as an Inmixing of an Otherness Prerequisite to Any Subject Whatever." Here, Lacan posits subjectivity as the outcome of a process modeled after "the real numerical genesis of the two" (2007, 191), and as an alternative to the unified, intentional consciousness of phenomenology. Lacan depends on principles of addition and repetition to construct the two, beginning with the empty set: "you have at one the place of zero and afterward it is easy to understand how the place of one becomes the second place which makes for two, three, and so on" (Lacan 2007, 191). Lacan's reference is to Frege's principles of construction from zero.[2] The process of generating natural numbers continues simply by repeating the operation of addition. However, Lacan (2007, 191) attributes to "the mark," or the empty set, "the effect of rubbing out the difference," and in doing so "makes it possible to add things with no consideration as to their differences." In the Lacanian accounting, the procession toward the two occurs with no regard to the unique status of the elements being combined, a demand prescribed by the "emptiness" of the sequence's point of initiation.

Deleuze and Guattari propose a reversal to the pattern suggested by traditional materialism by shifting their attention to a multiplicity constituting the unity from which the two emerges. Traditional materialism, they claim, "has never understood multiplicity" (*R* 51), since multiplicty exceeds an ideal and orderly production (and re-production) of reflection. Deleuze and Guattari are occupied, not with assessing the periodization of materialist thought but rather with its reimagining. They counter the *book-as-imitation* with a vision of the book as "radicle-system or fasciculated root" (*R* 51), a system articulated according to its own logic and thus jettisoning attachment to reflection, reproduction, and adequacy. Deleuze and Guattari's system *resists* unification though unification may be a consequence or accident; multiplicity describes a structure or system of organization and one that develops in spite of the absence of a principal root.

The system "of elaborate secondary roots" Deleuze and Guattari align with multiplicity depends on connection. Deleuze and Guattari thus name the *rhizome* as their preferred system, a system evident in animal life ("warrens, in all their functions as habitat, provision, passage, evasion and disappearance" [*R* 52]), and in botanical life, as an "underground stem-system from those roots or radicles" (*R* 52). This system develops apart from any logical procession, and as such is not beholden to evaluation of its adequacy. It is instead assessed according to six "approximate characteristics": *connection, heterogeneity, multiplicity, asignifying rupture, cartography, and tracing-mania*. As the third of the "approximate characteristics" of a rhizome, Deleuze and Guattari posit their version of multiplicity as a way of breaking the image on which materialist ontology depends. They are clear that multiplicity must be "treated as a substantive" (*R* 53). To treat multiplicity as a substantive is an acknowledgement of the *being* of the system; it means acknowledging that whatever *is* is fundamentally organized as a self-constructing and self-regulating system.

Deleuze and Guattari posit their rhizomatic multiplicity against the book as imitative unity, growing from a single point of origin and elaborating "in one direction, a linear one for example, while all the more affirming a totalizing unity in another dimension, that of a circle or of a cycle" (*R* 52). The circle or cycle restricts connections, training their proliferation in one direction. In the example I offered from Lacan above, development of subjectivity proceeds according to addition, a cycle repeated without attention to the differences or unique features of the subject being constituted. To avoid this simultaneous constraint and capitulation to unity, Deleuze and Guattari offer a procedural suggestion:

> In reality, it is not sufficient to say "Long live the multiple", even though this cry is difficult enough to utter. No typographical, lexical, or even syntactic trick will suffice to make it audible. The multiple *must be made*; not by always adding a further dimension but, on the contrary, in the simplest way possible, by force of moderation, at the level of the dimensions at our disposal, always *n* minus one (it is only in this manner that the one forms part of the multiple, through being always subtracted). Subtract the unique from the multiplicity to be constituted; write to the power of *n*-1. (*R* 52)

Deleuze and Guattari suggest that the multiple is discovered by subtraction; the differences disguised by Lacan's method of addition are actively sought out in the subtractive process. Indeed, the preceding comparison of the *book-as-imitation* to the *book-as-system* should suggest why a claim like 'long live the multiple' poses a challenge to the traditional materialist. Since materialism requires a unifying factor immanent to the material conditions that will secure transformation—an element or group with a unified motivation emerging from the mass—a commitment to multiplicity that effectively *demotes* unity to an accidental feature of that multiple will be an unsatisfactory proposal. However, the claims concerning subtraction yield a result even more dissatisfying; the tendency toward unity is something that must be actively resisted in the work of thought. While a system will never be complete, Deleuze and Guattari suggest that any move in this direction returns to the problematic system of reflection that has in fact stagnated.

In "Rhizome," Deleuze and Guattari are occupied, not with assessing repetition and periodization of materialist thought, but rather with reimagining it entirely; their assertion of multiplicity is instrumental in their revision. Deleuze and Guattari explain that substantive multiplicity "loses all relationship to the One as subject or object, as natural or spiritual reality, as image and world" (*R* 53). Multiplicity jettisons the One and any requirements of adequacy that may attend its representation. Deleuze and Guattari explain that their multiplicity "has neither subject nor object, but only determinations, magnitudes, dimensions which cannot increase in number without changing in its nature" (*R* 53–54). The multiplicity, in other words, is only its structure, a structure that is also explicitly mathematical, here drawing on language of quantity from Aristotle and—most significantly—from Bernhard Riemann, whose innovations in non-Euclidean geometry provide Deleuze with the vocabulary and principles to construct his account. I discuss the mathematical underpinnings of this commitment in Chapter 3 and return to Deleuze and Guattari's claims concerning the principle of multiplicity in more detail.

1977: Potatoes, Cuts

So far, I have explained the critical position in Deleuze and Guattari's "Rhizome" as disposed against traditional materialist logic; Deleuze and Guattari reject the articulation of matter and structure captured in the phrase "the One that becomes two." Their criticisms give way to a positive expression of materialism as a multiplicity that resists unification. However, "Rhizome" is not immune from criticism; it precipitates several responses from Alain Badiou in 1977. The first, "The Flux and the Party," offers a thoroughgoing critique of the version of materialism Deleuze and Guattari offer; the second, "The Fascism of the Potato," follows the materialist critique of the first text. Here, Badiou—writing pseudonymously as Georges Peyrol—emphasizes the problematic nature of the structure Deleuze and Guattari propose, attended by an inadequate conception of multiplicity.

In 1977, the UCFML (*Union des Communistes de France Marxiste-Leniniste*), a group of thinkers persisting in their commitments to Maoism, published *La situation actuelle sur le front de la philosophe* in two installments. The first took Althusser as its target, and identified a problematic quietism concerning the pivotal events of May '68 and their effects in the theoretical and political landscape.[3] The second installment, written as a polemic against Deleuze and Guattari, contained two essays by Badiou. One, "The Flux and the Party: in the Margins of Anti-Oedipus," is primarily directed against Deleuze and Guattari's understanding of a mass as motivated by 'desire.' This motivation, as Badiou ([1977b] 2012, 179) understands it, is *extra*-rational insofar as it eludes rational or logical accounting and suffuses the mass. This leads *not* to rupture and revolution, but rather to a quiet petering-out of state organization and is thus emptied of transformative possibility. Badiou ([1977b] 2012, 181) explains, "the 'leftist' political daydream is a mass movement that proceeds straight on until it is joyfully proclaimed that the State has quietly faded away." In other words, not only are Deleuze and Guattari surprised by the events of May '68, since these cannot be rationally or logically accounted for, these events can only go out with a whimper; they can neither predict a revolutionary rupture, nor can their system harness it for revolutionary change. Notions of 'desire' and 'freedom' are thus too vague to move the mass toward transformation.

Accompanying "The Flux and the Party" in the second installment of *La situation actuelle sur le front de la philosophe* is "The Fascism of the Potato," an explicitly Maoist polemic written by "Georges Peyrol," a pseudonym of Alain Badiou's.[4] Peyrol's account begins with a retrospective rehearsal of the events of May '68 from the point of view of "its intellectual protagonists," for whom the events of May were simply a "mass uprising," an "insurrection of the multiple" (*FP* 191). Continuing a theme from "Flux and the Party," these "protagonists" minimize any possibility of unification—however temporary—of this mass under a common *class* element. Peyrol claims that destabilizing instruments of class consciousness (e.g., unions and the *Parti Communiste Française* [PCF]) created the illusion of an "horizontal storm," consisting in the mobilization of "immigrants, women, ecologists, soldiers, prisoners, students, homosexuals, etc." (*FP* 192). To obscure the class condition necessary for the novel political and procedural position to emerge, Peyrol explains that they inflamed the multiple "against the pretensions of the One" (*FP* 192). This resulted in ideological and practical dispersions, since the multiple as a mass could not effectively unite; Peyrol claims, the interests of the intelligentsia *prevented* it from doing so:

> Under the anti-organizational pretexts, it is not too difficult to see the rejection of the point of view of class. (The petty-bourgeois intelligentsia's) theme was to add up the revolts (immigrants, women, ecologists, soldiers, prisoners, students, homosexuals, etc.), to enumerate the punctual social forces to infinity, but obstinately to combat anything resembling the political unification of the people's camp, seized in its antagonistic inflection, in its living class being. (*FP* 192)

The intelligentsia thus worked effectively to add up the various fronts of insurrection, but avoided the site of unification, the insurrection's "living class being," that might bring the desired novel order and procedure into being. The political situation is clear, insofar as the intelligentsia mobilized the multiple, but prevented it from mobilizing itself using the condition of class, a condition Peyrol identifies as *immanent to* the mass uprising.

Peyrol explains, however, that this mobilization on its own was insufficient, saying "this revolt against the pseudo-centres was far from giving way on the spot to the new Maoist thought, which is that of a centre of a new type (of the party of a new type), new not only in its being but also in its process" (*FP* 191). The productive conclusion of the uprising requires both novel organization and novel practice, as opposed to simple enumeration and recognition; Peyrol proposes here Maoism, on account of its ability to "concentrate" this class point of view and focus it toward revolution and transformation (*FP* 192). This immanent class condition, "the One of the multiple," functions also as "the politics of the people," and the content, but not the form, of the revolt. To capitalize on "the One of the multiple" seems to emphasize the integrity of the class movement emerging in the context of the mass revolt. As Peyrol has explained, these "politics of the people" were not realized, and the result comes in the appearance of "the politics of their enemies: political history abhors the void" (*FP* 193). The upshot seems to be the mobilization of the multiple without a direction, a problem or vacuum that would have been filled except for the intelligentsia working to avoid militancy and class struggle (*FP* 193).

The "Fascism of the Potato" is not only a broad and retrospective analysis of conditions leading to the failure of May '68, however. It turns toward a specific critique of Deleuze and Guattari, and their proposals in "Rhizome" as thought emblematic of this failure. Peyrol's engagement with Deleuze and Guattari proceeds along two paths. One is an analysis of the *structure* of the multiple or multiplicity, while the other concerns the *procedure* by which a relation of One to this multiple/multiplicity is produced. He begins with procedure—attempting to resolve what he takes to be an erroneous analysis of the dialectic, and then turns to a rejection of the structure posed by Deleuze and Guattari's theory of multiplicity.

Peyrol begins with Deleuze and Guattari's apparent misunderstanding of the dialectic: "Only a moron (*seul un crétin*)," Peyrol explains, "can confuse the Marxist dialectical principle 'One divides into two' with the genealogy for family ties concealed in 'One becomes two'" (*FP* 194). Peyrol takes issue with their description of the dialectic as *becoming*, as if the two emerges or is generated from the One. He insists, instead, that "the One has no existence as entity," since for the Maoist "there is only unity from movement, all is process" (*FP* 194). The One is, in this context, the result of dialectical, not genetic, process. Peyrol re-describes the dialectic as *scission*, or cutting from the multiple a temporary unity (*FP* 194). In the principal case of May '68, it is the cutting of 'class' from the 'mass,' which is a tactic of dividing from the multiple resulting in a temporary unity; this unity is alone capable of transformation, of unifying the groups identified in the so-called "horizontal storm" toward transformation. The image of taproot and 'principal unity' that Deleuze and Guattari ascribe to Maoism is *inverted*, and the implication is that the Maoism to which Peyrol is committed sees the mass as that *from which* a unity is identified. Without the work of scission, he insists that an ontology premised on multiplicity alone results in "the flattest of conservatisms, the surest ratification of everything that exists" (*FP* 196). Absent a mobilizing antagonism, Peyrol argues that Deleuze and Guattari's rhizome may unfold, but it will never change in any meaningful way.

With scission-as-procedure the preferred understanding of the dialectic, Peyrol turns to the structure of multiplicity. He begins by taking up two of the principles Deleuze and Guattari deploy to articulate the rhizome's structure: the first principle of connection and the third principle of multiplicity. Peyrol rehearses Deleuze and Guattari's first claim concerning the rhizome: "Any point of a rhizome can be connected to anything other, and must be" (*R* 52–53, quoted in *FP* 197). Peyrol targets the 'must be' of Deleuze and Guattari's claim and articulates three senses of this imperative to connection. The first is connection without motivation; individuals connect for the sake of connecting, for "the enjoyment of unlimited contacts" (*FP* 197). Second, a politics of the pure multiple can result only in failure. He explains that "since at least the Commune these 'convergences' of dismembered struggles are the prequel to the failure, the massacre, and the restoration of the One in its most repugnant military forms" (*FP* 197). History seems to suggest that the presence of a "horizontal storm" is neither necessary, nor sufficient a condition to accomplish political transformation; in fact, it seems to make way for fascistic control—Peyrol urges the "Sectarians of the rhizome" to "remember Chile!" (*FP* 197). Peyrol's worry is that a flattening cannot resist authoritarian tendencies; the 1973 coup and ascendance of Augusto Pinochet in Chile are, per Peyrol, outcomes following from the adoption of a "flat" multiplicity.

A third critique is registered against the principle of connection, which Peyrol uses to reinforce the value of scission and dialectical procedure. Because the principle of connection prevents antagonism—everything has a place by its being connected to other elements of the rhizome—there are no unanticipated elements that would precipitate a revolution. He insists that the work of scission is to generate a One that "(concentrates), through the practice of antagonism, the means to radically *separate* the people's revolutionary politics from *all* forms of bourgeois politics" (*FP* 197). Peyrol insists, at regular intervals in this critique, that the One generated through scission is both *precarious* and *temporary*. It is precarious because it is a contingent result of scission and temporary insofar as it will eventually be re-absorbed into the mass from which it is drawn. This reabsorption is necessary to accomplish the work of periodization necessary to the movement of materialism. This is opposed, to be clear, to the persistence of a One realized as military junta, in which control of the mass is accomplished by one group, with more permanence and destructive potential than a temporary condition uniting the mass.

Peyrol's final turn in "The Fascism of the Potato" is toward Deleuze and Guattari's third principle of multiplicity. In "Rhizome," Deleuze and Guattari define a multiplicity as having "neither subject nor object, only determinations, magnitudes, and dimensions that cannot increase in number without the multiplicity changing in nature" (*R* 53–54, quoted in *FP* 198). Peyrol finds this definition unsatisfactory, first on account of its 'occult' nature (he calls it an "incantation" [*FP* 198]), and second because the claim of dimension and its transformation "lives as a parasite on the dialectic" (*FP* 198). Peyrol is committed to an account of change that proceeds from the procedure of scission; the change to any multiplicity is evidence of there having been a unity sufficient to transform it. However, Deleuze and Guattari seem to suggest that a multiplicity changes only when the number of connections increases. Peyrol insists that these changes *require* the multiple to be "a thinkable category only in its contradictory relation to the One. All thinking of the pure multiple carries like its shadow a thinking of the pure one" (*FP* 198). Deleuze and Guattari cannot, despite their attempts to the contrary, offer an account of pure multiplicity that succeeds in rejecting the One, since the One of the rabbit warren is also and already the One of matter. As far as Peyrol can tell, Deleuze and Guattari have not escaped the problematic oscillation between One and multiple that a concept of multiplicity is designed to supplant.

Indeed, it is in this context—Deleuze and Guattari "ensnared in the Greek traps of the One and the Multiple" (*FP* 199)—that Peyrol argues against the procedure by which "the multiple must be made" (cf. *R* 52): subtraction. Peyrol argues that this subtractive move is "a complete failure! The subtraction of the One merely metaphorizes the need for both the One and Multiple, both '*n*' and '1' in Deleuze and Guattari's construction of multiplicities" (*FP* 199). Peyrol's rejection continues the line of critique he raised against Deleuze and Guattari's notion of procedure. This subtractive move emphasizes "the ideas of the revolt" and "the forces of revolution" *minus* the very things that transform these from mere description to productive revolution. Subtraction, Peyrol claims, accomplishes a problematic division that makes the whole vulnerable to fascist control, and Deleuze and Guattari are named as "prefascist ideologues" (*FP* 201).[5] This charge, which Peyrol unhesitatingly levies at Deleuze and Guattari, is avoided in the procedures he and other Maoists affirm; the One resulting from scission is precarious enough that it will not become tyrannical, but instead will be replaced by a new (equally precarious, equally temporary) One. Put another way, the practice of scission and its status as a temporary one cut from a pre-existing mass avoids the threat of tyranny.

Each of these critiques, whether aimed at Deleuze and Guattari or at Deleuze alone, emphasizes the same set of issues: namely, that Deleuze and Guattari's version of materialism is too enamored with the 'mass' in 'mass uprising,' and as such will resist or prevent a mobilizing factor—namely class—from unifying the mass. As Badiou puts it in the December 1977 lecture "Answering—to the Sphinx—demands from the subject not to have to answer—for the Sphinx," Deleuze's emphasis on the mass is such that "every unified configuration is an illusion, or even a 'totalitarian' threat" (*TS* 206). Badiou describes Deleuze as committed to a heretical materialism, the "dynamicist heresy," which insists "that all is flux, tendency, approximation" (*TS* 206). In the examination of "Rhizome" above, I emphasized Deleuze and Guattari's rejection of relying on a unity to describe multiplicity; their multiplicity is always *moving*, *connecting*, *generating*, and so resists slowing in a way that would support a rupture or accommodate an antagonism. Put another way, the dynamicist heresy is *all* remainder and *no* number, *all* matter and *no* reflection, *all* articulation and *no* revolution.

Deleuze and Guattari prefer a multiplicity defined by the number of connections holding between 'nodes' or poles. The examples they use to illustrate this version of multiplicity suggest the multiplicity is *localized* or *focused*; rabbit warrens, stem-systems, and dandelions, each develop connections from somewhere and in relation to one another. Their multiplicity is 'made' by a subtractive procedure, one that removes or suspends an organizing unity that would mute the distinctive and unique connections comprising the manifold. Writing as Georges Peyrol, Badiou argues for a relation of One and multiple in which the One is "of the multiple" (cf. *FP* 192). Peyrol insists that *this* One, cut from the multiple, is precarious because its existence is temporary and contingent. More specifically, Peyrol's polemic insists that the Maoists alone (among the *après-Mai* rabble) could correctly identify the condition of class that would have transformed the mass revolts. However, these positive positions are embedded in critique. Deleuze and Guattari reject the Maoist dialectic and its structural apparatus, while Peyrol points to the practical failure of rhizomatic organization in a highly-charged political atmosphere.

Returning to Badiou's work in the late 1970s raises two critical signals. First, Badiou's critique of Deleuze (with or without Guattari) is as old as the publication of *Anti-Oedipus*; it is right to see in this early phase figures of what becomes repeated in *Deleuze: The Clamor of Being*. I have briefly discussed the polemics appearing in *La Situation Actuelle sur le front de la philosophie*, as well as Badiou's lectures of 1977–1978, collected in Part IV of *Theory of the Subject* where Deleuze is mentioned. This set of critiques identifies a second signal: While Badiou is, indeed, engaged in vociferous critique of Deleuze and Guattari, he accomplishes this alongside the positive articulation of a materialist ontology and epistemology. In the translator's introduction to *Theory of the Subject*, Bruno Bosteels (2012, xvii–xviii) explains that the lectures late in 1977 are primarily concerned with Badiou's commitment to understand the movement of materialism—its "periodization"—and the conditions required for its renewal. It is not that the critiques of Deleuze and Guattari are a sideline; the polemical force of the writing, particularly in *La Situation Actuelle* belies that concern. Rather, they are part of a much larger theoretical enterprise, at least from Badiou's point of view. The issue turns on the *structures that both support thought and keep it moving*; in the 1970s this issue exercised in against an explicitly political background. In the 1990s, the question shifts to the quality or constitution of the multiplicity from which thought is formed. It is this concern that occupies Deleuze and Guattari, in their 1991 evaluation of the ontological proposal Badiou offers in *Being and Event*.

1991: Logic and Multiplicity

In the preceding, I examined an early debate concerning multiplicity couched in terms of *structure* and *procedure*. By structure, each camp describes the nature of multiplicity or mass; by procedure, they describe methods by which the relation of One to multiplicity is accomplished. In "Example 12" of Deleuze and Guattari's 1991 *What is Philosophy?* they take up Badiou's ontological proposal according to both the nature of the multiple and the way Badiou's work with it proceeds. They are particularly concerned with his use of set-theoretic strategies that align with logic; in the context of their analysis, Deleuze and Guattari see this as a problematic position that transforms philosophy's concepts into scientific functions and burdens these with evaluation according to logical principles. I begin by describing Deleuze and Guattari's preferred framework, which aligns philosophy and the concept (on the one hand) and science with the function (on the other).

Philosophy, Concept, Science, Function, and Event

Deleuze and Guattari answer the titular question of *What is Philosophy*? by claiming, "philosophy is the discipline that involves creating concepts ... they must be invented, fabricated, or rather created and would be nothing without their creator's signature" (*WP* 5). In addition to its task (the creation of concepts) and its persona (the philosopher), Deleuze and Guattari explain that the place of philosophy is the plane of immanence, the site that "secures conceptual linkages with ever-increasing connections" (*WP* 37). In other words, the plane of immanence is the proper place of linking and multiplying; concepts take advantage of this by "populating the plane" (*WP* 37). Deleuze and Guattari can therefore assign the *what*, the *whom*, and the *where* of concept production. Their first engagement with these elements, found in "Example 1," is with "one of the best-known, signed philosophical concepts, that of the Cartesian cogito" (*WP* 24). Though I discuss this example in greater depth in Chapter 4, the analysis of the cogito offers helpful background for the approach Deleuze and Guattari take to Badiou's work.

As a concept, Descartes' cogito—expressed as an event in the phrase, "I think, therefore I am"—is constituted by elements of doubt, thought, and existence, and the relationships holding among these, in their particular Cartesian context. The concept is thus a map of the movement of thought between these elements, and these passages are themselves content-rich, as I demonstrate in my analysis of the diagram in Chapter 4. A philosophical analysis of the concept, by Deleuze and Guattari's lights, requires recognition and understanding of these elements and their connections in their particular fullness, and also in their *locality*, since the components any concept collects are forged in a complete and localized whole. Put another way, the cogito is a local unity of doubting, thinking, and being.

Just as philosophy's output is the concept, science's output and means of communication is the function. These, according to Deleuze and Guattari, "are presented as propositions in discursive systems," and provide a vocabulary that "enables the sciences to communicate" (*WP* 117). Unlike philosophical knowledge, scientific bodies of knowledge are communicated propositionally and as such depend on fixed frames of reference in which these propositions can be interpreted. In Deleuze and Guattari's analysis, functions rely on "a relation of dependence or correspondence (necessary reason) … so that 'being human' is not itself the function, but the value of f(a) for a variable x" (*WP* 135). In other words, the function reports relationships of input and outputs that are interpreted for a given function on account of its corresponding to (or not) the states of affairs it purports to describe. Fixed frames of reference—and a rigid correspondence relation—allow for the development of scientific disciplines and their discursive peculiarities, but their dependence on propositions allows for these discourses to be sensible in other scientific frames. Deleuze and Guattari explain that as long as science facilitates communication and common vocabulary among its discursive bodies, it takes as its site the *plane of reference*, populated by "states of affairs, objects or bodies, and lived states" (*WP* 150–151) all of which express functions using propositions.

Philosophy's site, the *plane of immanence*, is populated by events, which express concepts. For example, the designations cogito (the subject of "Example 1") and "the ox head wired for sound" (the Kantian concept at the heart of "Example 4") each express concepts that are complex linkages and maps of relations between elements—multiplicities. They are events insofar as they each mark a novel irruption of thought consolidated in a transformed way; the ox head—as a mapping of thought—is re-forged from components of thought, doubt, and being that constitute the cogito. It is this notion of the event that is especially at issue, since Deleuze and Guattari claim in "Example 12" that Badiou's version of the event, as presented in *Being and Event* depends on logical operations (with their origins in science) and not from philosophical ones.

Logic and Badiou

Logic capitalizes on the work of science, especially its propositional character, and applies a second, more rigid, system of reference for judging the truth or falsity of the proposition. In this transformation, Deleuze and Guattari explain that logic "is reductionist not accidentally but essentially and necessarily: following the route marked out by Frege and Russell, it wants to turn the concept into a function" (*WP* 135). This is a mistaken pathway, since logic treats concepts as instances of mere predication. Per Deleuze and Guattari:

> The set of a function's truth values that determine true affirmative propositions constitutes a concept's *extension*: the concept's objects occupy the place of variables or arguments of the propositional function for which the proposition is true, or its reference satisfied. Thus, the concept itself is the function for the set of objects that constitute its extension. In this sense, every complete concept is a set and has a determinate number; the concept's objects are the *elements* of the set. (*WP* 136)

In this context, a concept is a predicate. For example, the concept *human being* requires objects to instantiate it; the elements *B*, *N*, and *T* are among the set constituting this concept, and the concept thus 'extends' to include it. Propositions concerning *B*, *N*, and *T* and their status as human beings are thus true affirmative propositions. Under this analytical framework, concepts also have an *intension*, or "the conditions of reference that provide the limits or intervals into which a variable enters in a true proposition" (*WP* 136). Intension describes the attending features or differences of an instance as a mechanism for evaluating the instance's correspondence to a concept. If *B*, *N*, and *T* are instances of the concept or set *human being*, they can also be assigned as part of the set "Becky's nieces and nephews." This second, more particular assignation, assigns to these instances a further specificity. Deleuze and Guattari call the intensive features of a proposition its *prospects*. However, and problematically for Deleuze and Guattari, from the vantage of logic concepts are neither created, nor are these forged; rather, concepts are predicative placeholders that either apply or do not, containers waiting to be populated and intensified using subsidiary forms of predication.

Deleuze and Guattari say, "By confusing concepts with functions, logic acts as though science were already dealing with concepts or forming concepts of the first zone" (*WP* 140). Logic doubles and reinforces the scientific function, mistaking its results for "philosophical" content and subjecting the creative pathways of thought evinced in *philosophy* and its *concepts* to the rigorous demands of propositional reference. Put another way, in its attempts to bridge science and philosophy, logic submits concepts like cogito or "the ox head wired for sound" to a determination and evaluation by correspondence, a relation itself determined by a single frame of reference; this frame of reference assigns values of true and false to each instance. However, this yields some difficulty when approaching Descartes' concept: To evaluate the cogito on these terms, one would evaluate the claim "I think, therefore I am" according to its correspondence to the existing frame of reference, and then subject each of the interior elements—expressed propositionally as "I doubt," "I think," and "I exist"—to a similar type of evaluation. Though it is the case that key 'moments' of a concept can be expressed propositionally, not all of the concept's components are expressed in this way, and the interior elements are reduced to mere predication. What appears is an analysis of any concept such that only the propositions remain, while the passages that trace thought's movement among the elemental stops drop away. In other words, a logical analysis of a concept only attends to a small portion of its components, identifying the unique features of an instance under broad (and indistinct) parameters.

In "Example 12," Deleuze and Guattari criticize Badiou for building a philosophical system from scientific foundations. They say Badiou "proposes to distribute at intervals on an ascending line a series of factors passing from functions to concepts" (*WP* 151). It is the "passing from functions to concepts" part of the claim that identifies the problem: Badiou gets the order of things wrong, since his theory of the event is developed from the function rather than from a conceptual system. In short, the system in Badiou's work that would give rise to an event is built on a category mistake. This limitation is especially evident in Deleuze and Guattari's exposition of the philosophy-concept versus science-function relationship, which amounts to a comparison of types of multiplicities and an articulation of their appropriate results. Per Deleuze and Guattari, it is the "neutralized base, the set, which indicates any multiplicity whatever" (*WP* 152), which forms the point of departure for Badiou's procedure. By beginning from "any multiplicity whatever," Badiou's system depends on a kind of arbitrariness about the starting point, one in which "any old set" might do. This is problematic because sets and the events linked to these are not localized; they are insufficiently particular and effectively come out of nowhere. Deleuze and Guattari counter this with the insistence that a multiplicity requires not a single set or element, but "at least two multiplicities," because "the multiplicity is precisely what happens between the two" (*WP* 152). Defining multiplicity as a dynamic relation accomplishes the requisite work of localizing the multiplicity, at least as far as Deleuze and Guattari are concerned.

Their critique also emphasizes the set's subjection to "a regime of the 'counting-as-one' (bodies or objects, units of the situation)" (*WP* 151). As I pointed out above in the analysis of extension and intension, the concept "Becky's Nieces and Nephews" has at least three elements: *B*, *N*, and *T*. Affirmative propositions follow concerning *B*, *N*, and *T* and their relation to the predicate-concept, and thus their belonging to the set. There is no dynamism in this form of counting, no creation or sense of irruption that Deleuze and Guattari assign to the properly-sited (philosophical) concept; Badiou ascribes philosophical productivity to a function, a logical mechanism depending on extension and intension, rather than to the creative trace of the concept.

Though the theory of the event is nominally at issue, Deleuze and Guattari's criticisms are here directed at the structural commitments that underwrite Badiou's account. They take issue with reframing creative and dynamic movements of thought in propositions, and then developing these propositions in an evaluative and rigid frame of reference. There is no apparent attention to locality or specificity, only re-presentation of thought's movement in static terms. This critique recalls Deleuze and Guattari's rhizomatic rejection of the dialectic, which draws from an established unity and re-presents this unity. In the discussion of logic and its relationship to science, logic takes the basic commitment of scientific procedure—functions resulting in propositions—and freezes these results in a way that re-casts philosophy and local philosophical interventions. Absent a localizing anchor, Badiou's theory of the event becomes unsatisfactorily abstract and, perhaps, as much a surprise to Badiou as May '68 appeared to Deleuze and Guattari.

1997 (and Beyond): The Clamor over *Clamor*

The preceding two sections demonstrate that multiplicity is a live issue for reflection and interlocution in work by Badiou and Deleuze and Guattari, beginning in the late 1970s and continuing through the 1990s. The final entry in this conversation comes in 1997, with the publication of Badiou's book-length treatment of Deleuze's work, *Deleuze: la clameur de l'être* (*Deleuze: The Clamor of Being*). This text interprets

Deleuze's work thematically and with a particular objection in mind: Deleuze's work is "organized around a metaphysics of the One" (*DCB* 17). In Badiou's critique of Deleuze and Guattari's "Rhizome," which I discussed above, Badiou suggests that their refusal to reproduce or re-present matter in thought—in short, an outright *resistance* against one-ness—ends with Deleuze and Guattari accepting the monism of matter. If everything is generated from matter, then all there is *is* matter; no meaningful place is made for the creative activity or separation thought can accomplish. In the terms I've been using so far, this critique is aimed at *procedure*.

Badiou also lodges an explicit critique of structure in Deleuze's work. As Badiou recounts the historical shifts in their "nonrelationship" (*DCB* 1), he claims that his work came into direct alignment with Deleuze's after the publication of *L'Être et l'événement* in 1988. "In developing an ontology of the multiple," Badiou says, "it was vis-à-vis Deleuze and no one else that I was positioning my endeavor" (*DCB* 3). This direct philosophical alignment emerged after decades of indirect personal and political opposition, which Badiou chronicles in the text's early pages. Put another way, the account in *Clamor* is developed from a site of continuity (a shared commitment to multiplicity, such as this concept is expressed in their respective works), rather than from a site of antagonistic difference, apparent in Deleuze and Guattari's critique of Maoism, and in Badiou's/Peyrol's critiques of the intellectual left and Deleuze and Guattari's affinity with this position.

Of this direct alignment, Badiou identifies "two paradigms that govern the manner in which the multiple is thought," assigning to Deleuze's "the 'vital' (or 'animal') paradigm of open multiplicities (in the Bergsonian filiation," and to his own "the mathematical paradigm of sets, which can also be qualified as 'stellar' in Mallarmé's sense of the word" (*DCB* 3–4). Badiou's characterization, then, situates his multiplicity alone as mathematical, and although Badiou acknowledges Deleuze's mathematical interests (noting that in the earliest stages of Badiou's knowledge of Deleuze it is because of the latter's work with math and Spinoza), he does not see these as relevant to Deleuze's conception of multiplicity.

This analysis, outlined in the first chapter of *Deleuze: The Clamor of Being*, begins with the assertion, "we can therefore first state that one must carefully identify a metaphysics of the One in the work of Deleuze" (*DCB* 11). It would appear, for example, that in *Difference and Repetition* when Deleuze argues against the traditional opposition of One and many to multiplicity (cf. *DR* 182), he is working to escape the demands of this couple. Badiou claims, however, that "going beyond a static (quantitative) opposition *always turns out to involve the qualitative raising up of one of its terms*" (*DCB* 10; emphasis mine). By avoiding the demand of the pairing *One* or many, the implication is that both are: one and multiple—as the totality represented by all available options—is one. Badiou identifies these as weak claims concerning multiplicity in Deleuze's oeuvre. He cites claims of the univocity of Being in *Difference and Repetition* as ontologically more significant than "the whole thousand-voiced multiple" for which Being speaks (*DR* 304, quoted in *DCB* 11). From *The Logic of Sense*, Badiou cites the 'all' of communication between events, the claim that Sense "is 'the position in the void of all events in one'" (Deleuze 1990, 180, quoted in *DCB* 11). In these (early) cases, multiplicity gives way to One in Deleuze's work; Badiou reads these texts as expressing a telos of the One-All, in which multiplicity serves as a kind of smokescreen to disguise this tendency.[6]

Another angle of criticism Badiou raises against Deleuze's work concerns the place of the subject. Badiou approaches this from the *machine-automaton* pairing, explaining that for Deleuze, the subject has emerged from the machinic production of desire, and sees its own disintegration in its return to the machine. There is a form of ascetic purification to this subject. Badiou claims, "Thinking," the activity associated with the subject, "is not the spontaneous effusion of a personal capacity. It is the power, won only with the greatest difficulty against oneself, of being constrained to the world's play" (*DCB* 12). As an alternative to the distinctive work of subjectivity in terms of *thought* and *choice*, Deleuze's subject takes up its task while tethered to the machine that produces it. This position, clearly unpalatable to the theorist of the militant subject, apparently limits the transforming power a subject ought to have. There is no room for the revision of an accepted order. In fact, Badiou sees the Deleuzian proposal as fundamentally backwards; the machine order forms a limit determining the subject's thought. This constraint—and the occasional or rare possibility for a subject to escape it—suggests to Badiou that "Deleuze's conception of thought is profoundly aristocratic" (*DCB* 12), since only a few can pass from "only meager local configurations" of the machine, through "the transfixion and disintegration of their actuality by infinite virtuality" (*DCB* 13), or the limit of this configuration's power. The counterintuitive result, Badiou claims, is that a philosophy whose currency is a vitalist multiplicity is also a philosophy of death, when considered from the position of the subject. This cycle, rare as it is, does not accommodate transformation except of the subject by way of asceticism, in which " ... thinking consists precisely in ascetically attaining that point where the individual is transfixed by the impersonal exteriority that is equally his or her authentic being" (*DCB* 13). By this analysis, thought is an individual in contemplation, a position Badiou assigns to "ontologies of presence" in *Being and Event*, and one utterly devoid of the disruptive possibilities Badiou builds into his own theory of the multiple.

Badiou's third criticism of Deleuze consists in the claim that the latter "arrives at conceptual productions that I would unhesitatingly call *monotonous*, composing a very particular regime of emphasis or almost infinite repetition of a limited repertoire of concepts, as well as a virtuosic variation of names, under which what is thought remains essentially identical" (*DCB* 15). By this, Badiou suggests that the apparent diversity of cases from which Deleuze draws is employed in service of the same conceptual structure across his oeuvre. Where some readers see *consistency*, Badiou sees a problematic *conservatism*, presenting itself in a commitment to a system abstracted from the cases the system purports to explain. Badiou takes as his chief example here Deleuze's analysis of cinema, in which "(Deleuze's) entire enterprise is proposing a creative repetition of concepts and not an apprehension of the cinematic art as such" (*DCB* 16). In the *Cinema* volumes, hundreds of films stand as cases illustrating Deleuze's notions of movement and time; this analysis is not 'film theory' as such, but rather a theory of time, and a theory of movement. Badiou's issue here is with the *displacement* occurring between case and concept. He says, again in the instance of the *Cinema* texts, " ... it is once again, and always, (Deleuze's) philosophy that begins anew and that causes cinema to be there *where it cannot, of itself, be*" (*DCB* 16). The cases are drafted to reiterate conceptual commitments. Put another way, cinema (and any other set of cases put forward) is emptied of its *cinematic* significance and stands only as an expression of Deleuze's conceptual apparatus. Badiou's critique poses keen questions concerning the issue of method; I return to this in my discussion of *Clamor* in Chapter 5.

Badiou closes his opening chapter of *Clamor* by crystallizing the critique he's raised, lauding the Deleuzian commitments to "an ethics of thought that requires dispossession and asceticism," and to "a systematic and abstract" philosophy as virtues (*DCB* 17). The first commitment, that Deleuze's philosophical project "is organized around a metaphysics of the One," is the source of dispute and its assessment. Readers of this final entry in the conversation concerning multiplicity are aware that Badiou offers an incomplete characterization of Deleuze's version of multiplicity, and from this characterization goes on to argue that Deleuze's commitments to univocity are liabilities for any purported claims to immanence. The objections present in Badiou's 1997 monograph rehearse once again the issues of structure and procedure present in the early work, and they also overlook the mathematic ideas that underwrite Deleuze's notion of multiplicity. This mode of reading Deleuze's work is not new; as I discussed above, in 1977 Georges Peyrol dismisses the third principle of multiplicity in "Rhizome" as an occult presentation or an "incantation". Badiou's work does not, however, go unremarked upon.

In addition to the immediate responses—and defenses–of Deleuze's overall program by Arnauld Villani and José Gil in *Futur Anterieur*, Daniel W. Smith ([2003] 2012) takes care to unpack each of the mathematical traditions on which Deleuze draws, thus responding to the unfulfilled promise of *Clamor*: An examination of the mathematical underpinnings of Deleuze's theory of multiplicities. In doing so, Smith rejects the circumscription of Deleuze's work to the so-called vitalist paradigm. Smith devotes an entire section to the broad and varied mathematical roots on which Deleuze's engagement draws, situating it primarily in a framework of problematics. This approach to mathematics involves, for example, the dynamic work of constructing a figure from a few rules for *drawing* and principles for deploying the tools, where a theorematic system *demonstrates* characteristics of an already-drawn and available figure (Smith [2003] 2012, 290–291); the difference in these sorts of systems turns on whether one *creates* (as in a problematic approach) or *consults* (in a theorematic or axiomatic approach). Deleuze favors the former, though not to the exclusion of the latter. Smith ([2003] 2012) marks the presence of axiomatics, particularly in *A Thousand Plateaus*, as a way of affirming the ecumenism of Deleuze's overall approach.[7]

Inattention to Deleuze's emphasis on the problem, as Smith ([2003] 2012, 307) explains, "results in numerous infelicities in (Badiou's) reading of Deleuze." In particular, he cites Badiou's assertion that one must begin from "a concrete *case*" when discerning Deleuze's ontological work (*DCB* 14), and argues that this demand begins from the end and not the beginning; the concrete case expresses a solution to a problem or the answer to a question. When armed with a solution, as Deleuze explains in *Difference and Repetition*, it is easy to forget the problem that occasioned and thus assume—incorrectly—that the solution is the problem (*DR* 162–163). In this way, Smith responds to Badiou's claim in *Clamor* that Deleuze relies on 'monotonous productions,' or simply the re-production of his general structural concerns in diverse fields. However, as Smith—following Deleuze—suggests, it is not merely structural reproduction. Rather, each solution is unique to the problem occasioning it; put another way—and in terms, I develop in subsequent chapters—the system or the 'concrete case' actually describes a particular and peculiar relationship between multiplicity and the one expressing it.

In addition to reflection on the nature of the system Deleuze develops—the relation of structure and procedure—Smith's critique establishes a directional or dispositional element necessary for distinguishing Deleuze's program from Badiou's. Smith ([2003] 2012, 309) claims, "Deleuze's is a 'bottom up' ontology (from problematics to discretization-axiomatization), whereas Badiou's is a 'top-down' ontology (elaborated exclusively from the viewpoint of axiomatics, denying the existence of problematics)." This is the difference between an ontology understood *genetically* or *dynamically*, tasked with the work of creation and construction, and an ontology understood *descriptively* or *statically*, in which questions of being and its appropriate science—though comprehensive in their development—are stripped of their constructive ability and instead demonstrate or describe. In Chapter 5, I explicitly return to this insight and point out ways in which the specific critiques raised in the set of engagements I have described reinforce this directional difference.

Smith argues that an appreciation of Deleuze's use of mathematics counters Badiou's claims that Deleuze's multiplicity is only vitalist. This is accomplished by recognizing Deleuze's use of various other models, alongside the mathematic, as illustrative of the core concerns of differentiation, individuation, and transformation in Deleuze's work ([2003] 2012, 305). There are complex mathematical structures at play throughout Deleuze's oeuvre, though Badiou—in his narrow demands for what counts, not only as ontology, but as mathematically appropriate for ontology—fails to appreciate both their presence and the nature of their deployment.

To be sure, the conversation does not stop with Daniel W. Smith's assessment. As I pointed out in the preceding chapter, *Clamor* initiates an entire strain of scholarship intended, in some cases to defend a more developed account of Deleuze's broad commitments against Badiouian critique (Roffe 2012; Crockett 2013). Simon Duffy (2013) has carefully accounted for the mathematics at work across Deleuze's oeuvre, defending Deleuze against Badiou's critiques after a comprehensive airing of Deleuze's engagement with mathematical tools. These are the major English-language respondents to the challenges raised by Badiou in his monograph on Deleuze. My work in the next chapter takes a deep dive into the nature and structure of mathematical multiplicity as Deleuze and Badiou respectively develop it.

Conclusion

My aim in this chapter has been to situate Alain Badiou and Gilles Deleuze with Félix Guattari in conversation over the nature and function of multiplicity. This conversation, which I claim begins with the publication of Deleuze and Guattari's "Rhizome" in 1976 and ends with Badiou's 1997 monograph *Deleuze: The Clamor of Being*, indicates that their positive presentations of a theory of multiplicity are often staged as a critique of the other's position. I began with an early and explicitly political point of discussion, initiated by *Rhizome* and followed immediately by several responses from Badiou. I focused my attention on "The Fascism of the Potato," in which Badiou (writing as Georges Peyrol) objects generally to forms of materialism that refuse to ratify a temporary point of consolidation, and specifically to the the principles of multiplicity proffered by Deleuze and Guattari as promoting a kind of political conservatism and thus undermining any moves toward transformation from the political left.

While the discussion of multiplicity in the 1976–1977 exchanges proceeds under political auspices, the analysis Deleuze and Guattari take up in 1991's *What is Philosophy?* is conceptual. Deleuze and Guattari approach Badiou's theory of the event—which they discuss briefly in "Example 12"—in their analysis of logic. Logic is problematic for Deleuze and Guattari because it reduces concepts, the currency of philosophical creation, to a scientific language that is evaluated according to a fixed plane of reference. Deleuze and Guattari locate Badiou's theory in this logical context, noting that the multiplicity to which any event is connected is insufficiently localized and overly abstract. Badiou's theory depends on mechanisms of logic that effectively freeze the movement of elements, rather than allowing for their free circulation and connection, like those discussed in "Example 1" and "Example 4" earlier in *What is Philosophy?*

I indicated that Badiou's chief criticism appearing in *Deleuze: The Clamor of Being* is the claim that in spite of commitments to the contrary, Deleuze proffers a "metaphysics of the One," rather than a comprehensive theory of multiplicity. This position has problematic implications for the nature of connection in Deleuze's ontology: first, what appears to be a philosophy of life ends up being a philosophy of death, since any unique instance is re-appropriated by the multiplicity from which it emerged; second, robust and unique productions (e.g., cinema) are appropriated as mere cases of the structure on which Deleuze's theory depends. Though Deleuze maintains a consistent theoretical commitment across his work, Badiou points out—as he did in "The Fascism of the Potato"—that this is really a commitment to a kind of conservatism of the "same," underwritten by the nature of multiplicity itself.

These three engagements allow me to identify two valences of critique: the first, structure, attends to the way the multiplicity itself is constructed; the second, procedure, attends to the way any one emerges or is cut from the multiple. In the next chapter, I demonstrate the separation of their theories along the lines of their explicit mathematical commitments. Attention to mathematical paradigms of multiplicity and their use by Deleuze and Badiou clarifies the *structure* of multiplicity; it is an investigation, then, of structure that forms the scope of the next chapter.

Notes

1. François Dosse discusses the relationship of the New Philosophers to Deleuze and Guattari; see Dosse (2010, 362–381). For an account of their relationship to Badiou, see e.g., Bosteels (2009, xvii–xviii) and Badiou (2016, 1–19).

2. Sam Gillespie (2008) explains that Frege's deduction of zero depends on a failed definition of number as "everything equal to itself," where "equal to itself" is the defining property. Gillespie (2008, 50) says, "Unfortunately for Frege, 'everything' was not a number. But if the logic is reversed, and we come up with the property 'not equal to itself,' we have the first logical concept that subsumes a number. *No object* falls under the concept 'not equal to itself,' and to that lack of object, we can assign a number, zero." Badiou (2008, 22) addressees Frege's deduction of zero.
3. See Badiou ([1977a] 2012, 11): "Althusser and company are more radically nihilist in that, for them, quite plainly *nothing happened* in May '68."
4. Badiou's continued Maoist affiliation through the 1970s required the use of pseudonyms in political, academic, and cultural criticism. In an interview with Antoine de Baecque, Badiou explains that his use of pseudonyms was an attempt to secure personal and intellectual protection. As a known member of UCMFL, Badiou explains that he and others "were arrested, frisked, interrogated. We had to take certain precautions, and the pseudonyms helped us cover our tracks so that the guy sitting across from us at the police station wouldn't have a ready-made file on us with texts we'd written and be able to say 'You wrote this, this, and this … '" (Badiou 2013, 13). Bruno Bosteels (2011, 110–156) devotes an entire chapter of *Badiou and Politics* (Chapter 3, "One Divides into Two") to Badiou's relation to Maoism and his participation with the UCFML.
5. Peyrol claims that this appellation follows from their commitments to the "negation of morality, repudiation of antagonism, aestheticism of the multiple, which outside of itself, as its subtractive political condition and its indelible fascination, leaves in abeyance the One of the tyrant: one prepares for the kowtow, one is already bowing down" (*FP* 201).
6. Badiou offers a litany later in *Clamor* that reinforces these problematic commitments: "We can say that, for Deleuze, Being is formulated univocally as: One, virtual, inorganic life, immanence, the nonsensical donation of sense, pure duration, relation, eternal return, and the affirmation of chance. As for thinking, this is, for him, disjunctive synthesis and intuition, the casting of dice, the ascetic constraint of a case, and the force of memory" (*DCB* 78). In "Deleuze's Vitalist Ontology," published in 1998 shortly after *Clamor*, Badiou considers the reasons that "made Deleuze choose against ontological mathematicity in the end, and what made him choose the word 'Life' as Being's main name" (*CT* 62); these reasons mirror those unpacked in the early sections of *Clamor*. Deleuze's fundamental commitment to univocal being, particularly as it is expressed in pp. 35–42 of *Difference and Repetition* and subsequent focus on the 'between' as this univocal being's primary indicator are liabilities Badiou identifies in the structure of being as Deleuze proposes it. Badiou also

argues that Deleuze's commitment to univocity prevents any partition of being in favor of emphasizing the links that gave rise to the partition in the first place (*CT* 63–64), a criticism directed at procedural separation of any one from the multiplicity. With respect to the 'between,' Badiou reads this as a commitment to the "neutral movement of the Whole," found by examining the link between conditions for determination and the determined being, rather than the resulting poles (*CT* 65). Together, these lead Badiou to claim that in Deleuze's project, "there is no being of Being" (*CT* 66), no aspect of being qua being that is not captured by the movement organizing it. This movement is *life*, "which is thinkable as an in-between of the movements of actualization and virtualization" (*CT* 66). The structure Badiou reads in Deleuze's work is the strands holding each of the poles together, but this structure is—by Badiou's estimation—only *vital*, never *mathematical*.

7. Jon Roffe (2016) has recently criticized, with characteristic care and attention, the understanding of axiomatics by Deleuze and Guattari in *A Thousand Plateaus*, and gestures to the way this misguided understanding of axiomatics forms the background of analysis in "Example 10" and "Example 12" of *What is Philosophy?*

References

Badiou, Alain. 2000. *Deleuze: The Clamor of Being*. Translated by Louise Burchill. Theory Out of Bounds Series. Minneapolis, MN: University of Minnesota Press. Originally *Deleuze: la clameur de l'être*. Paris: Hachette, 1997. (Cited in text as *DCB*)

———. 2005. *Being and Event*. Translated by Oliver Feltham. London: Continuum. Originally *L'être et l'événement*. Paris: Éditions du Seuil, 1988. (Cited in text as *BE*)

———. 2006. *Briefings on Existence: A Short Treatise on Transitory Ontology*. Translated, edited, and with an Introduction by Norman Madarasz. Intersections: Philosophy and Critical Theory Series, edited by Rodolphe Gasché. Albany, NY: State University of New York Press. Originally *Court Traité d'ontologie transitoire*. Paris: Éditions du Seuil, 1998. (Cited in text as *CT*)

———. 2008. *Number and Numbers*. Translated by Robin Mackay. London: Polity. Originally *Le Nombre et lest nombres*. Paris: Éditions du Seuil, 1990.

———. 2009. *Theory of the Subject*. Translated and with an Introduction by Bruno Bosteels. London: Continuum. Originally *Théorie du sujet*. Paris: Éditions du Seuil, 1982. (Cited in text as *TS*)

———. [1977a] 2012. "The Current Situation on the Philosophical Front." In *The Adventure of French Philosophy*, edited and translated by Bruno Bosteels, 1–18. Originally *La Situation actuelle sur le front de la philosophie*, 5–19. Paris: François Maspero, 1977.

———. [1977b] 2012. "The Flux and the Party: In the Margins of *Anti-Oedipus*." In *The Adventure of French Philosophy*, edited and translated by Bruno Bosteels, 171–189. London: Verso. Originally "Le flux et le parti (dans les marges de l' *Anti-Œdipe*)." In *La situation actuelle sur le front de la philosophie*, 24–41. Paris: François Maspero, 1977.

———. 2012. "The Fascism of the Potato." In *The Adventure of French Philosophy*, edited and translated by Bruno Bosteels, 191–201. Originally Georges Peyrol. "Le Fascisme de la pomme de terre." In *La Situation actuelle sur le front de la philosophie*, 42–52. Paris: François Maspero, 1977. (Cited in text as *FP*)

———. 2013. "'Cinema has Given Me so Much': An Interview with Alain Badiou by Antoine de Baecque." *Cinema*, 1–20. London: Polity.

——— with Gilles Haéri. 2016. *In Praise of Mathematics*. Translated by Susan Spitzer. London: Polity. Originally *Éloge des mathematiques*. Paris: Flammarion, 2015.

Bosteels, Bruno. 2009. "Translator's Introduction." In *Theory of the Subject*, vii–xxxvii. London: Verso.

———. 2011. *Badiou and Politics*. Durham, NC: Duke University Press.

———. 2012. "Translator's Introduction." In *The Adventure of French Philosophy*, edited and translated by Bruno Bosteels, vii–lxiii. London: Verso.

Crockett, Clayton. 2013. *Deleuze Beyond Badiou: Ontology, Multiplicity, and Event*. New York: Columbia University Press.

Deleuze, Gilles. 1990. *The Logic of Sense*. Translated by Constantin Boundas. New York: Columbia University Press.

———. 1994. *Difference and Repetition*. Translated by Paul Patton. New York: Columbia University Press. Originally *Différence et repetition*. Paris: Presses Universitaires de France, 1968. (Cited in text as *DR*)

Deleuze, Gilles, and Félix Guattari. 1981. *Rhizome: An Introduction*. Translated by Paul Patton. *I & C* 8 (Spring): 49–71. Originally *Rhizome: Introduction*. Paris: Les Éditions de Minuit, 1976. (Cited in text as *R*)

———. 1994. *What is Philosophy?* Translated by Hugh Tomlinson and Graham Burchell. New York: Columbia University Press. Originally *Qu'est-ce que la philosophie?* Paris: Les Éditions de Minuit, 1991. (Cited in text as *WP*)

Dosse, François. 2010. *Gilles Deleuze and Félix Guattari: Intersecting Lives*. Translated by Deborah Glassman. New York: Columbia University Press. Originally *Gilles Deleuze et Félix Guattari, Biographie croisée*. Paris: Éditions La Decouverte, 2007.

Duffy, Simon. 2013. *Deleuze and the History of Mathematics: In Defense of the "New"*. Bloomsbury Studies in Continental Philosophy Series. London: Bloomsbury.

Gillespie, Sam. 2008. *The Mathematics of Novelty: Badiou's Minimalist Metaphysics*. Anamnesis. Melbourne: re.press.

Lacan, Jacques. 2007. "Of Structure as an Inmixing of an Otherness Prerequisite to Any Subject Whatever." In *The Structuralist Controversy: The Languages of Criticism and the Sciences of Man*, 186–200. Baltimore, MD: Johns Hopkins Press.

Roffe, Jon. 2012. *Badiou's Deleuze*. Montreal: McGill-Queen's University Press.

———. 2016. "Axiomatic Set Theory in the Work of Deleuze and Guattari." *Parrhesia* 23: 129–154. Available online: https://www.parrhesiajournal.org/parrhesia25/parrhesia25_roffe.pdf. Accessed 8 September 2017.

Smith, Daniel W. [2003] 2012. "Mathematics and the Theory of Multiplicities: Badiou and Deleuze Revisited." In *Essays on Deleuze*, 287–311. Edinburgh: Edinburgh University Press.

CHAPTER 3

Structure: Multiplicity and Multiple in Deleuze and Badiou

In the preceding chapter, I situated Badiou in conversation with Deleuze and Guattari in the period 1976–1997. The content of this conversation is their respective theories of multiplicity and the problems or liabilities one thinker sees in the work of the other. I used an early pseudonymous work of Badiou's—"The Fascism of the Potato," written as Georges Peyrol—to establish two valences of understanding multiplicity: one, procedure, concerns the relationship between any one and the multiple from which it is drawn; a second, structure, describes the nature of the multiplicity itself.

My focus in the present chapter is with structure, or the nature of multiplicity. In what follows, I reconstruct Deleuze's and Badiou's theories of multiplicity on the terms of their original presentations and with the mathematical influences on which they depend intact. The approach I take here and in the next chapter—effectively keeping their accounts apart from one another—allows for something of a full airing of the program without interruption by critiques from the other. I return to these critiques in this work's Chapter 5.

Since the preceding chapter closed with Badiou's effective 'last word' in the conversation, and in particular with the claim that Deleuze is a metaphysician of the One, I begin with Deleuze's theory of multiplicity. I focus on concepts Deleuze develops from Bernhard Riemann's 1854 "On the Hypotheses which Lie at the Bases of Geometry" (hereafter "Hypotheses"). Riemann's transformation of

B. Vartabedian, *Multiplicity and Ontology in Deleuze and Badiou*,
https://doi.org/10.1007/978-3-319-76837-3_3

continuous quantity from its Aristotelian presentation, the expression of continuous quantity in multiply-extended manifolds, and a resulting alteration in the comprehension of physical space each clarify Deleuze's theory of multiplicity. I demonstrate this by making the Riemannian contributions explicit Deleuze's definition of the virtual idea in *Difference and Repetition*'s chapter 4, and the analysis of smooth and striated space in Plateau "1440: The Smooth and the Striated" in *A Thousand Plateaus.*

I then turn to Badiou's multiple, beginning with a 'short course' in key terminology at work in his ontology, namely operation, presentation, situation, and structure. This offers a valuable preface for Badiou's multiple and its conceptual heritage in Georg Cantor's distinction of inconsistent and consistent multiplicities. Cantor develops this distinction as a solution to the Burali-Forti paradox, a set-of-all-sets paradox concerning ordinal numbers. I also discuss a significant tool in the form of Ernst Zermelo's axiom of separation, a mechanism for maintaining the separation between inconsistent and consistent multiples that emerges after Russell's paradox, perhaps the best-known of the set-of-all-sets paradoxes. As with my work on Deleuze and Deleuze with Guattari, I read Cantor's influence back through key texts in Badiou's ontological turn, namely portions of *Being and Event* and *Briefings on Existence.* I discuss the continuity this assertion of in/consistency provides for Badiou's program, from the concerns expressed in the theoretical and political critique of the 1970s to the ontological concerns from the late 1980s to the present. I then conclude with a brief word about the axiom of separation and its significance in bridging structure and procedure in Badiou's work, a prefatory gesture for my work in Chapter 4.

Multiplicities: Riemann, Deleuze, Deleuze–Guattari

In Chapter 1, I offered a brief explanation of the nature of multiplicity in Deleuze's work and its significance for shaping his ontology. Multiplicity replaces notions of being as One or many; multiplicity is the "true substantive, substance itself" (*DR* 182), indicating that even these consolidated notions of One and many are themselves multiplicities. In Chapter 2, I attended to multiplicity in Deleuze and Guattari's theory of the rhizome, which "has neither subject nor object, but only determinations, magnitudes, dimensions, which cannot increase in number without changing in its nature" (*R* 54). The rhizome-multiplicity resists placement in a proposition and the application of predicates from without to specify it; instead, it depends on interior transformations to the linkages comprising the rhizome. These are registered as changes to determination, magnitude, and dimension, and these index the Riemannian concepts on which Deleuze's account of multiplicity depends.

I engage the content of Riemann's "Hypotheses" to clarify the notions from this text that Deleuze deploys in his theory of multiplicity and show that an appreciation of Riemann's work clarifies Deleuze's use of multiplicity in relation to the virtual idea, to rhizome, and (later) to the concept. I begin by describing key features Riemann's "Hypotheses": his innovation with respect to continuous manifoldness, including the articulation of multiply-extended magnitudes, and the transformation this occasions for comprehending physical space. I then re-read two versions of Deleuze's multiplicity with these insights in tow: the virtual idea in *Difference and Repetition*; and multiplicity in Plateau "1440: The Smooth and the Striated."

Riemannian Ideas

Before launching into an assessment of Riemann's work, consider an example. The neighborhood in which I live is defined and 'contained' by cross streets, its boundaries inscribed on the Denver city map. It is also divided into several hundred parcels, measured according to their square footage and legal description. Each of these individual parcels is mapped, and survey measurements describe where one parcel begins and another ends; indeed, as mapping technology becomes more sophisticated, anyone searching for real estate listings can observe the shape and size of individual parcels on a given city street. My neighborhood, then, can be determined and defined using these discrete modes of measurement. However, this is only one way of defining a neighborhood. I understand the neighborhood by pathways between my house to Liz's next door, to Melissa's house two blocks over, and to Damien's house. There are routes to and from local watering holes, our gym, and the nearest grocery store; linkages between libraries, transit, and other neighborhood services are ready to be made. Surveys of neighborhoods include 'walkability indexes,' or measures of the degrees to which gentrification has taken hold, measures equally significant to understanding the nature of a neighborhood. My neighborhood, then, can be mapped with more depth and complexity than city grid's coordinates allow. In fact, and in this context, the grid is just a single case of mapping—a 'sheet' of information—that lay over a series of other maps or sheets.

It is the difference between narrow, numeric, or coordinate definitions of the neighborhood and its parcels on the one hand, and the complex relationships that give a neighborhood its sense on the other—that illustrates the transition and innovation Bernhard Riemann's "Hypotheses" offers to spatial thinking. This work begins with the claim that "geometry assumes, as things given, both the notion [*Begriff*] of space and the first principles of constructions in space" (Riemann 1873, 14). While geometry depends on both a concept of space and axiomatic principles to shape that space, Riemann is troubled by the absence of an adequate explanation linking these two givens, since geometers, mathematicians, and philosophers have each failed to sort it out. Riemann asserts that the gap between these two assumptions is closed by the development of "the general notion of multiply-extended magnitudes," one "entirely unworked" so far in the history of mathematics (1873, 14). He thus invents the bridge he sees necessary by working from concepts already available, namely those of magnitude, continuity and discontinuity, points and elements, measure and counting. The plan at the outset of the "Hypotheses," then, is to rebuild the foundations of geometry with concepts and basic logical relations.

This work of rebuilding consists in three tasks. First is Riemann's innovation of the "multiply-extended magnitude," a concept he builds out of "general notions of magnitude" (1873, 14). Second, he intends to show that the multiply-extended magnitude is "capable of different measure-relations, and consequently that space is only a particular case of a triply-extended magnitude" (Riemann 1873, 14). Finally, the work accomplished in the second part puts the manifold 'to work' in order, as Erhard Scholz (1992, 29) explains, "to improve the comprehension of physical space." I now turn to Part I of the "Hypotheses," and especially its Sections 1 and 2 to explain Riemann's reconstruction.

Riemann's emphasis on a conceptual mathematics is inaugurated in a procedural statement: "Magnitude-notions [*Grössenbegriffe*] are only possible where there is an antecedent general notion [*allegmeiner Begriff*] that admits of different specialisations [*verschiedene Bestimmungsweisen*]" (Riemann 1873, 15). Riemann acknowledges the dependence of his concepts of magnitude (*Grössenbegriffe*) on an already well-defined general concept. Aristotle's distinctions of continuous and discrete quantity, as these are presented in Part Six of the *Categories*, are the well-defined concepts in play here.[1] Lines, surfaces, and bodies are examples of continuous quantities, while numbers and syllable constitute discrete ones.

The distinguishing characteristic of discrete quantity is that its parts remain separate from one another. Aristotle illustrates this, in the case of number, with the examples of "two fives making ten," or of seven and three making ten. In these cases, the parts "do not join together at any common boundary but are separate" (Aristotle 1984, 4b28). Likewise, syllables *bi-* and *-shop* retain their distinct sounds independently of their combination in the word *bishop*. Further, it is possible to describe relationships between these parts. In the case of discrete quantities, a principle of "relative order" may apply; there is a discernible order of succession in counting, and a word dictates the order of its syllables. However, such an order is not always required—the parts 7 and 3 can be combined in any order to accomplish the result of 10.

By contrast, the continuous quantities can be broken into parts, and despite this break maintain "a common boundary at which its parts join together" (Aristotle 1984, 5a1-2). Aristotle assigns, for example, the point as a limit of a line. These quantities are related using a principle of relative position, determined by the relations of joined parts. In other words, continuous quantities can be 'located' by the positions of other parts. Aristotle (1984, 5a16-20) says, "The parts of a line have position in relation to one another: each of them is situated somewhere, and you could distinguish them and say where each is situated in the plane and which one of the other parts it joins on to." For example, given a line AB bisected at a point C, I describe segments AC and CB (thus distinguishing the parts), mark C as the point or common boundary at which these segments are joined, and locate segment AC based on the position of CB. Aristotle's distinction, when applied to the example of my own neighborhood, illustrates the difference between considering a neighborhood as surveyed coordinates (i.e., my lot takes up x number of square feet on defined parcel y) that carve the space into defined and incommensurate parts and considering it as a series of linked spaces—there is, for example, a ramshackle fence that describes my backyard and Liz's backyard next door. The fence serves as a 'common boundary' joining continuous quantities.

Riemann (1873, 15) explains that the sense of a magnitude is only available against the background of an "antecedent general notion which admits of different specialisations." Riemann describes these as either continuous or discrete types of manifoldness, using *Mannigfaltigkeit* to distinguish these from magnitudes. Riemann's work depends, first, on a distinction of magnitude from manifoldness, and a concept of manifoldness as *Mannigfaltigkeit*. In Chapter 1, I discussed the Kantian use of this term, noting that it refers to an indeterminate field of sense data (in the manifold of appearance) or as an indeterminate placeholder for an object of cognition, the object = *X*. *Mannigfaltigkeit* comes to Riemann by way of Gauss and Herbart; in Gauss's case, a manifold describes a "system of objects linked by some relations, the dimensionality of the manifold depending on properties and interconnectedness of the relations" (Ferreirós 2007, 44). Gauss's insight describes the 'interior structure' of a manifold, which (as I explain below) Riemann deploys in the "Hypotheses."

In tracing the passage of *Mannigfaltigkeit* from Kant to Riemann, Erhard Scholz (1992, 22) explains Riemann's dependence on a procedure from J.F. Herbart called "graded fusion," which consists in an activity of mind "which does not destroy the individual presentations but glues them together, with the result that the continuous transition from one to another becomes possible." Per Scholz (1992, 23), Herbart's graded fusion is crucial for Riemann's development of the manifold, since Riemann "presupposed the existence of concepts, mathematical or not, which may arise as the result of a 'graded fusion' into serial forms. He took up the result and opened it to mathematical consideration, thus forming his concept of *multiply-extended magnitude*." As Scholz explains here, and Ferreirós (2007, 46) and Biagioli (2016) confirm, a serial form is an abstract framework that, at least in Herbart's analysis, can be applied to and organize the content of the empirical world. For Riemann, Herbart's serial form offers some flexibility concerning the way any region of space behaves and thus establishes the way particular marks and relations in this space will operate and be measured. Put another way, where Gauss offers the interior structure of any manifoldness as comprised of relations, Herbart's insights describe the initial conceptual condition for the possibility of a manifold.

Riemann (1873, 15) says that continuous manifolds are developed from points and allow for a "continuous transition from one to another of these specializations," while discrete manifolds take elements as their units. Riemann (1873, 15) insists that for continuous magnitudes, "measure consists in the superposition of the magnitudes to be compared; it therefore requires a means of using one magnitude as the standard for another." The line AB bisected at C allows for the comparison of the individual segments AC, CB in relation to one another and in relation to the whole AB. If no "standard magnitude" is available, Riemann (1873, 15) says, "two magnitudes can only be compared when one is part of the other," and the comparison consists only in determining "the more or the less and not the how much." In other words, comparison of continuous magnitudes is a matter of relative position that may—in some cases—admit of basic quantitative evaluation (i.e., more or less).

Discrete manifolds are "so common that at least in the cultivated languages any things being given it is always possible to find a notion in which they are included" (Riemann 1873, 15). However, continuous manifolds are far less common; the most direct examples of multiply-extended manifolds are "the positions of perceived objects and colors" (Riemann 1873, 15). In the case of color, it is a set of relations holding between three values: grayscale (the deviation, either more or less, of temperature), gamut (the range of values that express the color), and luminance (that color's brightness). When, for example, Pantone releases a color portfolio, a hue—blue—is specialized as Aquamarine, Scuba Blue, and Classic Blue. These specializations are distinguished from one another depending on the values of grayscale, gamut, and luminance each one expresses.

Following the initial distinction of discrete and continuous manifolds, Riemann applies further specifications to these manifolds. He explains the difference between simply-extended continuous manifolds and multiply-extended continuous manifolds. Two continuous manifolds are simply-extended when it is possible to transition from one point to another (Riemann 1873, 15). This transition is essentially one path with two directions, backward and forward. Extension, then, describes the number of pathways available between points. In his articulation of the virtual idea as multiplicity, Deleuze uses terminology of dimension in place of extension; a simply-extended manifold is one-dimensional, while a discrete manifold has dimension zero (Plotnitsky 2009, 123). However, as Riemann (1873, 15) explains, a simply-extended manifold "...passes over into another entirely different, and again in a definite way, namely so that each point passes into a definite point of the other, then all the modes of determination so obtained form a doubly-extended manifoldness." It is therefore possible to add pathways by increasing the available

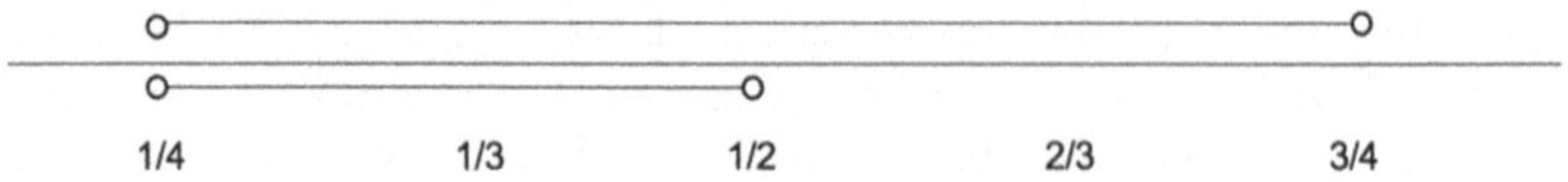

Fig. 3.1 Ordinary comparison of interval (1/4, 3/4) with interval (1/4, 1/2)

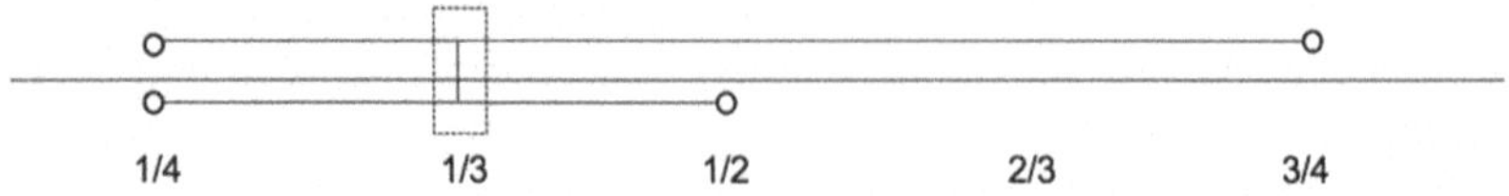

Fig. 3.2 Reading intervals (1/4, 3/4) and (1/4, 1/2) for shared subspaces

paths to travel. As Riemann describes it here, the simply-extended manifold is doubled by adding additional points and thereby creating further pathways for travel. This process continues to triply-extended, quadruply-extended, and to *n*-extended manifolds, or in Riemann's (1873, 15) terms, "as a composition of a variability of $n + 1$ dimensions."

Arkady Plotnitsky explains the process of extending manifolds using open intervals and the neighborhoods they constitute. He suggests that some continuous object (e.g., a line) may be described, "not by the set of its points, but by a class of its open subspaces covering it" (Plotnitsky 2009, 120). Take, for example, the two open intervals Plotnitsky proposes, (1/4, 3/4) and (1/4, 1/2). On visual inspection of Fig. 3.1, the magnitude described in (1/4, 1/2) is shorter than the magnitude described in (1/4, 3/4); the magnitudes are separate entities.

However, and as Plotnitsky suggests, re-reading this figure for *neighborhoods* requires a recognition that both the open intervals pass over the point 1/3 (Fig. 3.2). The open subspaces 'covering' the point 1/3 include both intervals, which form the neighborhood of subspaces for point 1/3 (Plotnitsky 2009, 120). The basic graphs of these intervals are therefore not compared or described by their length, but rather on account of their relationship at the point 1/3, a re-appraisal of space according to sites of overlap between two intervals instead of comparing these according to their (discrete) units of length; this is akin to comparing my neighbor's yard to my own using the boundary created by our shared fence, rather than calculating lot sizes using Denver city planning data.

The insights of dimensionality are also illustrated in the differences between the two diagrams. Figure 3.1 indicates two, one-dimensional intervals; there is only one available pathway to travel (i.e., from 1/4 to 3/4 and back). When the intervals are interpreted as overlapping at 1/3, the number of available pathways increases to four. These pathways include each of the one-dimensional pathways between the original intervals (so two pathways); a third pathway between 1/4 on the lower line through the point 1/3–3/4 on the upper line; and a fourth route from 1/4 on the upper line, through the point 1/3–1/2 on the lower line.

In Plotnitsky's example, the identification of neighborhoods and subspaces depends on a way of 'interpreting' manifolds. The differences between these very basic Figs. 3.1 and 3.2 reflect this shift. If the former Fig. 3.1 demonstrates discrete manifoldness, the latter Fig. 3.2 changes the interpretation of these intervals, expanding the kinds of determinations available by relationships holding between points, between intervals, or between other spaces. According to Plotnitsky (2009, 121n21), this expansion to open subspaces and neighborhoods "enabled Riemann to define manifolds of any dimensions, even infinite-dimensional ones, as collections of covering maps," and secured these manifolds to be determined independently of "ambient Euclidean space." However, Riemannian space and Euclidean space are not mutually exclusive. Plotnitsky (2009, 119) explains that Riemannian space is a more "capacious" concept; while a space or surface may be a complex, non-Euclidean space there are local neighborhoods that can be described in Euclidean terms.

Merzbach and Boyer (2011, 472) compare Riemann's differential geometry to "ordinary geometry," which "is interested in the totality of a given diagram or figure, whereas differential geometry concentrates on the properties of a curve or a surface in the immediate neighborhood of a point on the curve or surface." The comparison of Figs. 3.1 and 3.2 illustrates the differences between so-called "ordinary geometry," expressed by Fig. 3.1, and differential geometry, expressed in the shift to shared subspaces in Fig. 3.2. Returning to the example with which I opened this section, the various ways I understand the neighborhood in which I live do not eliminate its description as parcels and survey coordinates, but rather recognizes it as a method of definition *when* a certain vocabulary or disposition toward the space applies.

Reading Riemann in the Virtual Idea

In the preceding, I discussed three key ideas appearing in Riemann's "Hypotheses": his reiteration and expansion of continuous quantity as manifoldness, the procedure by which these manifolds become multiply-extended, and the consequences of these innovations for the comprehension of physical space. Principles of continuity, extension (or dimensionality), neighborhoods, and overlapping subspaces each offer mechanisms for emphasizing the relationships between spaces; this emphasis is significant for Deleuze's theory of multiplicity, which I return to below.

In Chapter 1, I gestured to Deleuze's assertion of the virtual idea as a multiplicity. I stated Deleuze's preference for an account of the idea using internal genesis rather an external conditioning; in the latter, time functions as a third term facilitating communication between categories and intuitions. By contrast, internal genesis indicates an idea emerges directly from, and can be traced to, the conditions that produce it. The virtual idea, discussed in chapter 4 of *Difference and Repetition*, is an instrument of internal genesis and depends on a structure afforded by a continuous multiplicity; this ground allows for the three pieces of the differential—the element *dx*, its reciprocal relation *dy/dx*, and its subsequent determination resulting in a pre-individual singularity—to take hold.

Deleuze is clear: "Ideas are multiplicities: every idea is a multiplicity or a variety" (*DR* 182). He indicates that any idea is issued from a multiplicity; put another way, the idea is structured according to the principles attending any multiplicity. However, Deleuze turns in on the very problem of being itself, noting that "Everything is a multiplicity in so far as it incarnates an Idea. Even the many is a multiplicity; even the one is a multiplicity" (*DR* 182). In his hands, the foundational dilemma of ontology finds its solution in a third option, and the calcified positions of One and many are re-animated as dynamic multiplicities. Instead of raising the stakes by confirming the many qua multiplicity as the solution, Deleuze takes a deflationary step; as solutions, one and many are themselves expressions of differentiated multiplicities.

These (apparently) colossally opposed positions are really separated by differences characteristic of the multiplicities to which they are tethered. "Everywhere," Deleuze says, "the differences between multiplicities and the differences within multiplicities replace schematic and crude oppositions ... there is only the variety of multiplicity—in other words, difference" (*DR* 182). The method of evaluation Deleuze proposes here is in line with the turn, discussed in the previous section, of Riemannian manifolds inward and their determinations as fundamentally local; this is opposed to the measure of any manifold according to a fixed background space. To understand the way the multiplicity develops these differences, an appreciation of the explicitly Riemannian terminology Deleuze deploys is valuable. Deleuze specifies the nature of the idea, describing it as "an n-dimensional, continuous, defined multiplicity" (*DR* 182). This definition demonstrates that whatever expression is under investigation, the multiplicity it expresses is structured according to basic Riemannian commitments: dimension, continuity, locality.

When describing dimensionality of a multiplicity, Deleuze explains, "by dimensions, we mean the variables or coordinates upon which a phenomenon depends" (*DR* 182). Riemann asserts that the continuous manifold can be either simply- or multiply-extended, based on the number of available transitions or pathways passing through a point. In the case of my own neighborhood, the dimensions can indicate the number of pathways etched between my house and various landmarks in the neighborhood. Deleuze uses the language of *n*-dimensionality to indicate the openness of the virtual idea according to multiple extensions and dimensions; in other words, it is not confined or restricted to a particular number of dimensions. Dimensionality equips the virtual idea to manage the complexity apparent in the multiplicity it expresses. Beyond the pathways holding between my house and landmarks, the dimensionality of my neighborhood increases to layers that include quality of life measures, transit routes, and other opportunities of connection that increase the complexity surrounding my house.

In terms of continuity, Deleuze claims this consists in "the set of relations between changes" holding among dimensions (*DR* 182). This is, more or less, the definition of a continuous manifold and also the conditions for its determination. That is, in the exposition of Riemann's account above I demonstrated that continuous manifolds are measured using relations developed in and by the manifold itself. This is a prescription of measure, which ensures continuous manifolds are not subjected to an evaluative framework outside the manifold itself. To submit the manifold to an exterior measure is to dilute or cover over the differences and specificity that makes the continuous multiple what it is. In the case of Riemannian manifolds, this exterior measure is ambient and static space; with respect to Deleuze's virtual idea, the exterior measure consists in demands of representation and identity.

Finally, Deleuze insists that the multiplicity is definite. He says, "by definition, we mean the elements reciprocally determined by these relations, elements which cannot change unless the multiplicity changes its order and its metric" (*DR* 182–183). This characteristic depends on the points linked by the multiplicity and emphasizes its locality. A definite multiplicity captures certain portions of "the intense world of differences" with which Deleuze's transcendental empiricism operates. It occupies part of this 'intense world' and reads for the relationships in that part.

Riemann's conceptual-mathematical approach identifies the principal framework of the virtual idea. That is, continuous multiplicity and its features provide an inchoate solution to the problem of external conditioning that occasions Deleuze's inquiry in the first place. By its very nature, continuous multiplicity is keen to an account of internal genesis because the linkages proper to that multiplicity are managed within it; there is no 'without' to the multiplicity, and thus no threat of exterior conditioning. The analysis of dimensionality in a Riemannian multiplicity identifies a complex and contained site from which an idea might arise, and its emphasis on locality or its status as definite focuses on a portion of space, rather than the space against a larger background or in a global configuration.

The broader transformation in ways of thinking about space, a hallmark of non-Euclidean geometries (in general) and Riemann's work (in particular), emphasizes not only immanent measure but a kind of ecumenism about measure in general. Merzbach and Boyer (2011, 497) explain that in investigating "distances between two points that are infinitesimally close together," Riemann insisted that there were more options available than traditional Euclidean methods, though there are cases in which traditional forms of measure may be activated. Deleuze's comments at the opening of this section in *Difference and Repetition*, in which he asserts the priority of multiplicity to one and many, offer a similar commitment; just as static background space might be invoked to describe the nature of a multiplicity, Deleuze insists that the complex series of connections underwriting a virtual idea are anchors from which the idea and its empirical expression may emerge. Where Riemann's work emphasized a shift in the comprehension of physical space, Deleuze's assertions with respect to multiplicity suggest a shift in the comprehension of ontological space. Immediately following the exposition of the virtual idea in *Difference and Repetition*, Deleuze offers a three 'cases' in which the virtual idea is operative: first, "atomism as a physical idea" (*DR* 184); second the notion of the organism as a "biological idea" (*DR* 184–185); and third, an analysis of Marxism as a social idea (*DR* 186).[2]

The third of these examples, the social Idea "in a Marxist sense" (*DR* 186), requires a radical localization for its prosecution. In this example, Deleuze explains that Althusser and his collaborators were right to reject an historicist interpretation of Marxism,

> since this structure never acts transitively, following the order of succession in time; rather, it acts by incarnating its varieties in diverse societies and by accounting for the simultaneity of all the relations and terms which, each time and in each case, constitute the present. (*DR* 186)

Deleuze here suggests that *if* there is a social Idea in a Marxist sense, it is not an idea that is rehearsed historically and in fact, does not repeat itself 'successively.' That is, any social Idea is borne in the present and is limited to that particular present. For example, the social Idea organizing relationships in my Denver neighborhood in the mid-1990s is not the same social Idea working out in Denver in the late 2010s. If, for example, the social Idea is something like 'gentrification,' the structure installs relationships between notions of 'opportunity,' 'affordability,' 'displacement,' 'property tax,' and 'growth,' as each of these are interpreted at present. I discuss this example again in the next chapter, but for the present purposes it transforms an idea of Denver away from the physical boundaries associated with it on a map and toward a series of related elements describing it in late 2017.

Finding Riemann in Smooth and Striated Space

A similar approach attends Deleuze and Guattari's discussion of Riemannian multiplicity in *A Thousand Plateaus*. There, Riemann's work is offered as one among several models for thinking about the relation of smooth space, a space of virtuality or ideality, and striated space, a space of actuality and spatio-temporal inscription. The models in question include a technological model, by which practices of weaving fabric are expressed; a musical model, explained using Boulez's distinctions of "smooth space-time," which "one occupies without counting" and "striated space-time" where "one counts to occupy" (*TP* 477); the maritime model, which illustrates smooth space with "intensities, wind and noise, forces and sonorous and tactile qualities, as in the desert, steppe, or ice" (*TP* 479) and the striated space with maps and bearings; a physical model demonstrating the passage of striated back to smooth space; and an artistic model (*TP* 493–494). Deleuze and Guattari identify two modes of relationship between smooth and striated space, the relation underwriting each of these models. The first is a distinction of magnitude from distance, and a second is the relation of distances to one another; the Riemannian principles I've outlined above assist in understanding these elements.

A magnitude describes a comparison determined against ambient space and is identified as a metric space; on the other hand, distance describes comparisons determined using those things being compared, and is nonmetric. Of distance, Deleuze and Guattari say that it is "a set of ordered differences, in other words, differences that are enveloped in one another in such a way that it is possible to judge which is larger or smaller, but not their exact magnitude" (*TP* 483). The judgment of 'larger' or 'smaller' is only available according to the distances being compared, and not according to fixed measure. This description is consonant with Riemann's descriptions of continuous multiplicities; I expressed this difference in my explanation of the Figs. 3.1 and 3.2. Deleuze and Guattari attribute to distances (and, by implication, continuous multiplicity) "a process of continuous variation" (*TP* 483), a recognition of the links developed and developing among points in the multiplicity. By contrast, "multiplicities of 'magnitude' distribute constants and variables" (*TP* 483); the measurements of magnitude accord more neatly with Euclidean spaces, rather than Riemannian ones.

While the first aspect concerns the relationship of metric to nonmetric space, a second aspect concerns the relationship of nonmetric spaces to one another. For this assessment, Deleuze and Guattari quote Albert Lautman's *Les schémas de structure*, in which Lautman raises first the fact that "Riemann spaces are devoid of any kind of homogeneity" (quoted in *TP* 485).[3] In other words, Riemann spaces are defined in such a way that they do not lose their individual determinations when connected to a larger space. Rather, they require principles of assembly—terms like gluing, stitching, or patching describe the connections of manifolds to one another. Lautman says, "Riemann space at its most general thus presents itself as an amorphous collection of pieces that are juxtaposed but not attached to each other" (quoted in *TP* 485). Consider, for example, the various ways in which I make qualitative descriptions about my neighborhood. The walkability indexes and trends concerning gentrification are metrics separate from the paths my husband and I take to our neighborhood restaurants, but these different metrics still describe the same portion of space; as such, they form layers or strata, and suggest a complex network of connections available for assessing the neighborhood in which we live.

Summary

In the preceding, I have identified and explained two examples in which attention to Riemannian concepts clarifies both the presence of multiplicity and its organization in Deleuze's and Deleuze-Guattari's work. These emphasize the nature of connection or relation as definitive of the Idea, and smooth space. Further, the disposition toward understanding space proffered in Riemann's work encourages Deleuze or Deleuze-Guattari's reader to see particularity against a series of other connections, other nodes for linking up, other expressions of intensity.

In Chapter 1, I mentioned Deleuze and Guattari's definition of the rhizome-multiplicity as "only determinations, magnitudes, dimensions which cannot increase in number without changing its nature" (*R* 53–54). Having rehearsed key Riemannian concepts, the nature of this rhizome-multiplicity is considerably clearer; the rhizome is its pathways, its connections, which are described dynamically and variably according to $n + 1$ dimensions. Like the rabbit hole on the surface, or the description of a neighborhood by parcel and survey coordinates, these determinations give way to a complex series of connections that underwrite and inform these points of entry. What remains is the task, as Deleuze and Guattari explain in Rhizome, of 'making' the multiple, that work of subtraction that identifies layers and analyzes the strata. This work, also aided by a Riemannian procedure, is the subject of Chapter 4. I turn my immediate attention to the structure of multiplicity in Badiou's work, which depends on an entirely different mathematical paradigm.

Multiplicities: Cantor, Badiou

In the preceding section, I discussed Riemann's contributions to geometry and comprehension of physical space and explained the ways these contributions illuminate the structure of multiplicity in Deleuze's individual work and in his collaboration with Guattari. The focus of the present section is to assess the mathematic sense of multiplicity in Badiou's work, drawing on his deployment of a distinction from Georg Cantor: that of inconsistent and consistent multiplicities, the distinction that founds Badiou's ontological project in *Being and Event*.

I begin, as I did in the preceding, by setting out the primary mathematical presentation; in this case, it is a set-of-all-sets problem posed by the Burali-Forti Paradox and Cantor's subsequent response. I then return to Badiou's work, focusing on Badiou's reading of Cantor and the distinction of inconsistent and consistent multiplicities in *Being and Event*. I discuss the way it shapes the framework for both an understanding of being qua being and the ontology Badiou develops in the text. After discussing the initial presentation of inconsistent multiplicity in *Being and Event*, I turn to a more detailed meditation on the concept in *Briefings on Existence*; in its opening essay, "The Question of Being Today," Badiou elaborates five features of inconsistent multiplicity that support a turn to mathematized ontology. This offers a framework for appreciating the deployment of the axiom of separation in Badiou's ontology, a principle regulating the relationship between inconsistent multiplicity and any consistent multiplicity constructed from it. I discuss the axiom of separation in its original context, following the presentation of Russell's paradox, and explain its function in Badiou's work. It is against this background that I discuss Badiou's break with Cantor over the notion of inconsistent multiplicity as an Absolute, a totality that only God can count. Badiou, by contrast, insists that inconsistent multiplicity is *no-thing*, though it is named by the void.

Cantor's Inconsistent and Consistent Multiples

In the Introduction to *Being and Event*, Badiou explains that the possibility of a mathematized ontology rests on the emergence of set theory as "a singular science," which itself depends on Georg Cantor's innovations with respect to understanding infinity (*BE* 6). In Cantor's 1899 letter to Richard Dedekind, he offers the conceptual separation of inconsistent from consistent multiples. The letter is precipitated by a problem raised by Cesare Burali-Forti two years earlier regarding the construction of a set of all ordinal numbers.

Burali-Forti's paradox and Cantor's solution each turn on the concept of well-orderedness, or well-defined multiplicity, and whether such a concept is only expressed in sets. The immediate upshot is, by Cantor's deduction, that the concept of well-orderedness may apply to any multiplicity but it is not necessarily the case that every multiplicity is a set. In other words, and as Jean van Heijenoort (1977, 113) explains, Cantor's solution to the paradox is "to abandon, not the well-ordering of the multiplicity, but its sethood." The Burali-Forti paradox concerns ordinal numbers, which (most basically) describes the position of an object or number in a sequence, (e.g., first, second, third). Burali-Forti (1977, 111) identifies a well-ordered set as having a first element, and every element "that has successors has an immediate successor." Take, for example, two collections or sets of ordinals, viz. (first, second, third) and (fourth, fifth, sixth); these sets satisfy the criteria for being well-ordered. In these example sets, each one has a first element [(first) and (fourth), respectively], and there are no gaps between the successor elements in each set. Arranging two sets sequentially indexes to each set its own ordinal: the set of ordinals (first, second, third) comes first in the sequence, while the set of ordinals (fourth, fifth, sixth) comes second.

The problem giving rise to the paradox, per van Heijenoort (1977, 113), is a contradiction "engendered by the ordinal of the multiplicity of all ordinals." This contradiction does not arise from the satisfaction of criteria for well-orderedness per se, but rather from attempting to describe the sets themselves as First and Second. The problem emerges because the descriptor ordinal First is greater than the ordinals in the segment (first, second, third); but (first) is also in First, such that (first) < First, which is a contradiction.[4] Van Heijenoort (1977, 104) puts it succinctly, saying "the ordinal is at once an element of the set of ordinals and greater than any ordinal in the set." The formulation of the Burali-Forti paradox shows that assigning sethood to ordinality constrains it in such a way that it results in a contradiction. Though my discussion concerns simple ordinals, the stakes of both the Burali-Forti paradox and Cantor's reply concern the logical possibilities of transfinite numbers. That is, if it is not possible to consistently describe simple ordinal numbers, then the prospects for transfinite numbers—numbers designating quantities larger than the natural whole (counting) numbers—are compromised.

In his letter to Richard Dedekind, Cantor takes up the contradiction posed by the sethood of ordinality. He agrees to the condition of well-orderedness, saying that "A multiplicity is said to be well-ordered if it satisfies the condition that every submultiplicity of it has a first element; I call such a multiplicity a 'sequence' for short" (Cantor 1977, 114). Cantor's definition condenses the requirements for succession in the terms of 'sequentiality,' and does so in the language of multiplicity; simply by recasting the problem in these terms, Cantor accomplishes an important conceptual addition.

Cantor (1977, 114) identifies a consistent multiplicity as a multiplicity whose elements "can be thought of without contradiction as 'being together,' so that they can be gathered together into one thing." Consistent multiplicity is equivalent to a set, and Cantor requires consistent multiplicity to hold without contradiction, its elements unified and complete. On these terms, the consistent multiplicity reads parallel to any set-language invoked by Burali-Forti. However, consistent multiplicity allows for the inference of inconsistent multiplicity, a label Cantor applies to a multiple for which "the assumption that all of its elements 'are together' leads to a contradiction" (1977, 114).[5] An inconsistent multiple is never closed, and by its nature will never constitute a complete unity. Cantor's division reveals a careful thinking through multiplicity in terms of contradiction; that is, while Burali-Forti identifies the relationship for which a contradiction holds, Cantor imagines a construction for which a contradiction may be avoided.

The distinction between consistent and inconsistent multiplicities allows Cantor to solve the Burali-Forti paradox. Cantor accomplishes this by explaining that the system of all numbers (Ω) is a simply-ordered system. A simply-ordered system designates a "rank order" holding between elements in a set, "such that, for any two of its elements, one is the earlier and the other the later" (Cantor 1977, 114).[6] Furthermore, as the system of all numbers, Ω contains "a least number. Hence the system Ω, when naturally ordered according to magnitude, forms a sequence" (Cantor 1977, 115). That there is a least number in the sequence is an insight Badiou will capitalize on in his discussion of the void in Meditations four and five of *Being and Event*, but for the immediate analysis this conclusion indicates that Ω (and the sequence beginning with zero, which Cantor describes as Ω') are well-ordered multiples, but they cannot be consistent.

If Ω' (the sequence of ordinals that begins with zero as its least number) was a consistent multiple, that is, if it could be completed as a set, Cantor (1977, 115) says, "there would correspond to it a number ∂ greater than all numbers of the system Ω." The set Ω would thus be assigned a corresponding number. The trouble with this is, however, that since ∂ is itself a number (the number corresponding to the supposed set of all ordinal numbers qua consistent multiple), ∂ "occurs in the system Ω, because this system contains all numbers; ∂ would thus be greater than ∂, which is a contradiction" (Cantor 1899/1977, 115). On this analysis, Cantor has walked right up to the edge of the Burali-Forti paradox. Without the distinction between inconsistent and consistent multiples, Cantor would be forced into retreat, but his subsequent claim is, "The system Ω of all numbers is an inconsistent, absolutely infinite multiplicity" (1899/1977, 115). There is, in other words, no set of all ordinal numbers, even though ordinal numbers may be conceived as belonging to a system—the ordinal number system.

Cantor is not yet finished, however. The claim that the system of all ordinal numbers is an inconsistent multiplicity (and thus incomplete and inconceivable as a unity) applies to transfinite cardinal numbers as well. To put it another way, the question of ordinality must somehow extend to Cantor's system of Alephs ($\aleph_0$, $\aleph_1$ ... $\aleph_n$). Cantor (1977, 114) explains that similarity between two systems consists of a "one-to-one relation such that the rank order of corresponding elements is the same in both." Where Cantor designated the system of ordinal numbers by omega (Ω), he identifies the system of all transfinite cardinal numbers by tau (ת). He insists that there is a "one-to-one" relationship between Ω and ת, such that these systems can be read (imprecisely) using simpler finite ordinality and cardinality; first, second, third corresponds one-for-one with 1, 2, 3. Thus, Cantor (1977, 116) claims that ת is "similar to the system Ω, and therefore likewise inconsistent; or absolutely infinite." In other words, one-to-one correspondence allows the transfer of a mapping strategy used for simple ordinality onto more complex systems.

Cantor's solution to the Burali-Forti paradox begins from a shared definition of well-ordered sets that relies on requirements of a first element and sequentiality. Cantor's solution departs, however, in its recasting the problem in terms of multiplicity, rather than sets. This allows Cantor to distinguish two types of multiplicity: consistent (a multiple for which the consequences of sethood does not yield a contradiction) and inconsistent (a multiple for which the closure of its elements into a set results into a contradiction). In addition to thinking types of multiplicity more broadly, Cantor's solution takes the step of assigning logical possibilities to these multiplicities, imagining 'both sides' using the framework of contradiction. From this terminological and inferential basis, Cantor argues:

1. The system of all numbers Ω is a simply-ordered system, which has a least number and is ordered sequentially;
2. And Ω cannot be a consistent multiple, since any ordinal ∂ used to designate it would be simultaneously in the system and greater than the system.
3. Therefore, Ω must be an inconsistent multiple; because of this inconsistency it must also be absolutely infinite.

In Cantor's solution, two clear innovations emerge. First is his distinction of types of multiplicity, which allows for a more capacious understanding of the nature of ordinals. Second is the insistence on developing this distinction with the simple requirements of contradiction. Inconsistent multiplicity can never be closed or counted as a totality, and Cantor's deduction of the inconsistent multiple secures against the contradiction that devastates 'mere' sets. Inconsistent multiplicity remains open and, from Cantor's position, absolutely infinite. These basic insights inform a reading of Badiou's version of in/consistent multiples, a reading that expands our understanding of Badiou's notions of presentation and operation. It is to this analysis that I now turn.

Badiou's Inconsistent Multiple

Cantor's distinction of multiplicity into inconsistent and consistent types clarifies Badiou's uses of these concepts. In the present section, I am interested in Badiou's descriptions of inconsistent multiplicity: first, as these appear in the early stages of *Being and Event;* and second, the expansion of this concept in *Briefings on Existence.* Though these two textual encounters show Badiou's account to be continuous with Cantor's, I conclude this discussion by explaining the break

Badiou makes with Cantor over the status of inconsistent multiplicity as an absolute; this break ushers in a brief engagement with the void and void set, the element that allows for the construction of consistent multiples. Attention to inconsistent multiplicity here foregrounds my work with consistent multiples in the next chapter, since it is consistent multiplicity that offers the material for understanding procedure in Badiou's work.

However, the initial appearance of inconsistent and consistent multiples in Badiou's work is not technical, but rather *conceptual* insofar as it appears in Badiou's offering of a shorthand that he deploys in mathematical, meta-ontological, and textual meditations in *Being and Event* (*BE* 19). This shorthand consists in the notions of *operation, presentation, situation*, and *structure*. Presentation and operation describe the landscape of being and its science respectively, while situation and structure describe the after-effects of an operation. For the purposes of the present examination, I focus on operation and presentation, which re-situate the problem of One and many as *one* and *multiple* in Badiou's work. This is a meta-ontological positioning; that is the operation-presentation couple allows Badiou the conceptual bridge between a traditional philosophical problem (the status of being qua being as one, many, or multiple) and the technical valence of presentation qua multiple that allows Badiou to initiate his peculiar ontological proposal.

In keeping with Badiou's decision against the One as being qua being, he situates the operation as the work of counting, that procedure for closing any multiple as a consistent set. Operation takes multiple as its material, but this connection is only determinable *retrospectively*, or as an "after-effect of the count" (*BE* 24). This retrospective identification of multiple secures the link of one and multiple necessary for any count whatever, at least conceptually. Identifying operation as the mechanism for one-formation in Badiou's ontology allows him to offer the conceptual pairing of *situation* and *structure*. Situation, Badiou explains, is "any presented multiplicity" (*BE* 24), a site at which multiplicity is exposed to the operation; it is the "place of taking place" for ontology. Structure, by contrast, indicates the presence of an operation; it is the evidence of the count-as-one having operated in any situation (*BE* 24). Without delving too deeply into this pairing (at least not yet, anyway), the articulation of terms like situation and structure allows Badiou to introduce the most basic relation of his ontology: *belonging*. He says, "when anything is counted in a situation, all this means is that it belongs to the situation in the mode particular to the effects of the situation's structure" (*BE* 24). Here belonging is belonging *simpliciter*, without the technical support of the axiom of separation that Badiou introduces in Meditation 3.

The triplet of operation-situation-structure offers a meta-ontological mapping for what Cantor identifies as a consistent multiple. Its mention here clarifies the nature of consistent multiplicity in the presence of a count-operation (situation) and belonging-relation (structure). The operation-situation-structure triplet also offers a means for attending to *presentation*, by describing the relation of a situation to the multiplicity it presents, insofar as that "exposed" multiplicity belongs to a situation. Though Badiou says, "the count-as-one (the structure) installs the universal pertinence of the one/multiple couple for any situation" (*BE* 24), this linking does not specify the nature of the multiple-presentation. Operation and presentation, though not interchangeable, are inseparable. Put another way, if there is an operation (closure by a count), there is also presentation (a multiple having been counted).

If the operation is the work of one-*ing*, presentation is aligned with the multiple. However, the *nature* of this multiple is not yet clear. Following the 'founding activity' of operation, Badiou explains the way multiple functions in both operational and "presentational" contexts:

> The multiple evidently splits apart here: 'multiple' is indeed said of presentation, in that it is retroactively apprehended as non-one as soon as being-one is a result. Yet 'multiple' is also said of the composition of the count, that is, the multiple as 'several-ones' counted by the action of structure. There is the multiplicity of inertia, that of presentation, and there is also the multiplicity of composition which is that of number and the effect of structure. Let's agree to term the first *inconsistent multiplicity* and the second *consistent multiplicity*. (*BE* 25)

Badiou assigns the Cantorian distinction on a fault line emerging in his account after operation. That is, there are two senses of multiple: the count is described as a multiple—a *consistent multiple* and is a multiple accompanying one insofar as any one is a composition of multiple elements; there is also a multiple preceding the count, the inconsistent multiple. Perhaps one of the more important identifications Badiou makes here concerns inconsistent multiplicity as a multiple of inertia; presentation does not mark a site at which being qua being shows itself or makes itself available to the operation; there is no active principle or movement associated with being qua being. Inconsistent multiplicity is a static, infinite reservoir situated at the foundation of ontological operation.

Badiou insists that in its preliminary stages, presentation's "only predicate" is "that of the multiple" and the sense of multiple here can be split according to consistent and inconsistent types (*BE* 28). Consistent multiples isolate specific instances of presentation: "the multiple of presentation is *this* multiple whose terms let themselves be numbered on the basis of the law that is structure (the count-as-one)" (*BE* 28). By contrast, inconsistent multiplicity is always and irreducibly multiple—it will never be constrained, closed, or counted. Badiou describes presentation in inconsistent multiplicity as "more latent ... a kind of inert irreducibility of the presented-multiple to appear" (*BE* 28). Badiou's assertion of the irreducibility of the presented multiple reinforces 'the multiplicity of the multiple' at the level of inconsistency; it is *general*, as an alternative to the apparent *specificity* of consistent multiples, and resists closure because of its inconsistency "all the way down."

Here Badiou's use of Cantor's distinction describes the landscape and limitations of ontology. This is to say that the project of *Being and Event* builds from this initial positing of both inconsistent and consistent multiplicity (on the one hand), and inscribing inconsistent multiplicity of ontological activity (on the other). In the next chapter, I discuss in detail the 'limitation' posed to consistent multiplicity by inconsistency. Now, however, I turn to a more detailed analysis of inconsistent multiplicity in Badiou's *Briefings on Existence*.

Briefings on Existence is a collection of essays in what Badiou identifies as "transitory ontology," or a phase in which the movement from *being* to *appearing* is still unfolding. In the first of these essays, "The Question of Being Today," Badiou offers a meditation on the nature of inconsistent multiplicity, identifying "five conditions for any ontology of the multiple" required to keep the One at bay (*CT* 40). These conditions align with certain of Cantor's observations of inconsistent multiplicity and are thus extended and interpreted in a Badiouian context.

First, Badiou explains, "Ontology is the thought of the inconsistent manifold, that is, of what is reduced without an immanent unification to the sole predicate of its multiplicity" (*CT* 40). This claim functions first as a statement of what *ontology* is (as the thought of inconsistent multiplicity); and second, as a description of inconsistent multiplicity itself.

Here, he claims that the inconsistent manifold resists unification *except* in the case of describing it as a multiplicity or manifold (i.e., the inconsistent multiple is multiple). This description is in accord with Cantor's insistence that the inconsistent multiple remain open, an insistence he offers as a remedy to the contradiction attending sethood. Of course, with respect to the first interpretation (an identification of what ontology is), this insistence is entirely Badiou's.

Second, Badiou identifies this ontology of the multiple as depending on the view that "the multiple is radically without-One in that it itself consists only of multiples. What there is, or the exposure to the thinkable of what there is under the sole requirement of the 'there is,' are multiple of multiples" (*CT* 40). To avoid the trap of transcendence, Badiou's claim that the "multiple is radically without-One" leaves the multiple as the frontier of ontological inquiry; there is nothing beyond the multiple of multiples, no possibility for unification in some other deeper 'sector' of multiplicity.

These first two assertions demonstrate Badiou's interpretation of his inconsistent multiplicity as aligned with Cantor's version. In the third assertion, Badiou expands the ways in which is view is aligned with that of Cantor; this expansion is in the terms of the philosophical problem of One and many, with which Badiou's ontology is (most generally) concerned. He says:

> Granted that no immanent limit related to the One determines multiplicity as such, there is no first principle of finitude. The multiple can thus be considered infinite. Or even, infinity is another name of multiplicity as such. As no first principle binds infinity to the One, it ought to be tenable for there to be an infinite amount of infinities, an infinite dissemination of infinite multiplicities. (*CT* 40)

Ontology does not, as Badiou says here, begin from a 'first principle of finitude.' This would circumscribe the multiple using some expression of one-ness; Badiou rejects phenomenology (and particularly that of Heidegger), on account of phenomenology's beginning from and limit according to Dasein's finitude. Absent this determining limit, Badiou can claim that the multiple is infinite and that the two concepts are identical. Cantor (1977, 114) affirms this in his letter to Dedekind, describing inconsistent multiplicity as "absolutely infinite." Badiou's subsequent claims regarding the 'infinity of infinities' are posited without reference to Cantor's absolute. As such, Badiou is required to maintain the inconsistent multiple as absolutely *open*, without a horizon of closure afforded by a transcendent presence (*CT* 55). Paul Livingston (2012, 53–54) explains that this transcendent presence underwrites a "transcendent orientation" of thought, which "sets up the totality of beings by reference to a privileged being, a 'super-existence' that assures the place of everything else, while at the same time obscuring its own moment of institution or the grounds of its own authority." This super-existence fixes the totality of beings, and in doing so forecloses against the possibility for change.

This third condition, along with the fourth ("Instead, let us consider a multiple to be a multiple of nothing" [*CT* 40]), identifies significant claims regarding not only the nature of inconsistent multiplicity, but also claims to existence. Recall that the notion of the 'there is' is restricted to the operation or count-procedure, and does not operate in being qua being. This requirement underwrites the fifth condition Badiou raises in *Briefings on Existence*, that "actual ontological presentation is necessarily axiomatic" (*CT* 40). That is, once a link can be established between being qua being and a first 'existence,' the rules governing this existence are axioms of set theory. In the next chapter, I explain the empty set, its status as first existent, and its relation to inert, inconsistent multiplicity. However, these claims depend on an articulation, first, of Zermelo's axiom of separation, and second an address of Badiou's break with Cantor over the nature of inconsistent multiplicity as absolute.

Breaking with Cantor: Separation and Comprehension

The discussion of inconsistent multiplicity from *Briefings on Existence* offers a second statement of its nature, particularly vis-à-vis the work of ontology. It 're-sets' the founding notion of Badiou's ontology in explicit Cantorian terms and disposes it toward the ontological practice of count-construction. In Cantor's case, inconsistent multiplicity is posited to keep ordinality open. Sam Gillespie explains, using Badiou's work in *Number and Numbers*, why Cantor requires a distinction of ordinals from natural numbers. As Gillespie (2008) points out, while natural numbers are ordinal numbers, not all ordinal numbers are natural numbers (for example, ω [the first transfinite ordinal], or $\aleph_0$). Gillespie (2008, 55) calls on Badiou here to point out that "an ordinal is the number of the figure of well-ordering," or that an ordinal is an index of well-ordering. As such, the 'options' for these sorts of indices must remain open, rather than restricted to the natural numbers.

The axiom of separation, proposed by Ernst Zermelo in 1908, regulates the relation between sets and the multiple from which these are drawn and, by implication, reinforces the persisting openness of inconsistency in Badiou's project. The axiom of separation reads, "Whenever the propositional function $G(x)$ is definite for all members of a set M, M possesses a subset M_G containing precisely those elements x of M for which $G(x)$ is true" (Zermelo 1977, 202). There are three operative components here: first, a function $G(x)$, an instrument or tool 'doing' the separating, which is also definite, meaning that when it is applied a result will obtain; second, an already-available set of elements M; and third, a subset M_G resulting from the application of the function to M. The subset "Becky's Nieces and Nephews" (a set I call on frequently in this text) results from applying a definite function—*x is a child of my sibling*—to the set "Becky's Family" or, more to the scale prescribed by inconsistent multiplicity, "human beings." The operation prescribed by the axiom seems a pretty straightforward mechanism for separating subsets from sets; it also telegraphs a relationship between inconsistent and consistent multiples, insofar as one can conceive of consistent multiples being cut from a larger—infinite—multiplicity.

It serves another important purpose, however: Zermelo's axiom of separation is a *limited comprehension principle*; any function operative on a set draws out a subset of the original, and does not define the entire set M. Significantly, Zermelo's axiom preserves the possibility for excess in the original multiple. As a limited comprehension principle, the axiom of separation is the mathematical procedure that protects against totalization or a destructive contradiction. Zermelo's axiom proceeds from his insights that "sets may never be independently defined by means of this axiom but it must always be separated as subsets from sets already given: thus contradictory notions such as the 'set-of-all-sets' or the 'set of all ordinal numbers' ... are excluded" (Zermelo 1977, 202). Zermelo's axiom is a procedure that secures set-theoretic operation against paradoxes that would undermine the possibility for set construction.

The paradox in question, at least in the development of axiomatic set theory, is Russell's Paradox, another set-of-all-sets paradox that emerges in response to Gottlob Frege's efforts to remove guesswork from certain of the presentations of naïve set theory. As Badiou points out, and as the discussion of Cantor's work above reinforces, the articulation of key notions in naïve set theory is conceptual and not properly mathematical; Badiou cites Cantor's initial definition of *set* as "the grouping into a totality of quite distinct objects of our intuition into thought" (quoted in *BE* 38) as one such example. Frege's work, particularly in the *Begriffschrift* presents a formal logical language in which these ideas can be expressed.

However, at issue is Frege's "Basic Law V," introduced in *Grundgesetze der Arithmetik*, which describes equality between two sets, each organized according to concepts, based on a one-to-one correspondence between the contents of the two sets. Paul Livingston (2012, 23) offers an example in which sets organized according to the predicates *red* and *heavier than 20 kg* may each "ensure the existence of a set containing all and only things" determined by these predicates; the set of red kettlebells in my gym and the set of kettlebells heavier than 20 kg in said gym each have the same number of objects *and the very same objects* populating these sets. As such, and according to Basic Law V, these sets are equal.

The problem with Frege's Basic Law V is that it is intended as a *universal comprehension principle*, describing rules to which all sets conform. As Bertrand Russell pointed out, not all sets are the same. In articulating his famous paradox, Russell initiates thinking about a *limit* on comprehension (put one way) or a rejection of totality (put another way). In his 1902 letter to Frege, Russell (1977, 125) states the inconsistency as follows: "Let *w* be the predicate: *to be a predicate that cannot be predicated of itself.* Can *w* be predicated of itself? From each answer, the opposite follows." Russell is essentially asking whether a particular predicate can both describe a set *w* and be included in the set *w*. The formal presentation—in keeping with the Fregean impulse toward clarity of expression—sets this out a bit more precisely:

1. R: $\{x \mid x \notin x\}$
   First, define the Russell set R as a set such that it does not belong to itself. The negated belonging symbol indicates this limitation.
2. $R \in R \leftrightarrow R \notin R$
   Second, infer from the definition of the Russell set R that if R belongs to itself, then it is the case that R *does not* belong to itself (reading the implication from left to right); but the double arrow

requires the implication be also read from right to left. As such, if *R* does not belong to itself, then *R* belongs to itself.

Russell introduces a contradiction that undermines the coherence and power of naïve formal language; per Badiou, it shows the "equivalence of a statement and its negation," which "annihilates the logical consistency of the language" (*BE* 41). To preserve consistency, then, a limit on comprehension must be established. Zermelo's axiom of separation allows for set construction without positing a required one-to-one mapping between the set and the multiple from which it is drawn.

As a limited comprehension principle, one that separates sets from an already-given multiple and thus avoids problematic set-of-all-sets consequences, the axiom of separation authorizes the work of separating subsets from any presented multiple. In the context of Badiou's project, the application of the axiom of separation must always be read according to the controlling problematic of his ontology: the multiple *is*, the one *is not*, and the excess of inconsistent multiplicity must be preserved by any ontological apparatus. Badiou's closing remarks in Meditation Three reinforce this requirement: "the axiom of separation takes a stand within ontology," since it reinforces the demand that a multiple must "be already-there; some pure multiple, as multiple of multiples, must be presented in order for the rule to then separate some consistent multiplicity" (*BE* 47–48). The axiom of separation suggests that if there is a multiple, it is possible to draw from it elements to form a set. This suggestion, however, is hypothetical until some parameter for existence can be established; the axiom of separation waits for the emergence of 'something' to count.

The axiom of separation functions on the side of structure in Badiou's ontological project, since Zermelo's axiom authorizes the limited comprehension of any set in relation to the multiple from which it is drawn. It maintains of excess on the side of inconsistency such that it will not be subject to closure once any count-procedure is inaugurated. This authorizes Badiou's rejection of Cantor's characterization of inconsistent multiplicity as *absolute*, and his own assertions that the inconsistent multiple is *nothing*.

Breaking with Cantor: Absolute or Nothing

Cantor's commitment to—or obsession with—infinity as a pathway to divinity is well known. Paul Livingston notes that the questions leading to Cantor's exposition of sets are motivated by "the thought of a divine, infinite transcendence." Livingston (2012, 22) explains this disposition toward divinity

> ... led Cantor to believe both that the well-defined infinite sets of the naturals, or the reals *can* exist as wholes (in that God can indeed group them all together as unified sets, even if finite agents cannot) and, on the other hand, that the whole infinite hierarchy of infinite sets forms an "unincreasable" ultimate totality that cannot be treated mathematically at all, what Cantor (following tradition) called the *Absolute*.

Livingston's comments make clear that as far as humans can count, infinity is infinite and cannot be totalized by the operations of organization or counting; this openness makes space for the *transfinite*, again as far as human counting goes. Beyond the transfinite, Cantor posits the presence of an "unincreasable totality" that constitutes the Absolute.[7]

Cantor's disposition toward an Absolute or transcendent conception of infinity marks the place from which Badiou's alignment with Cantor breaks. In Meditation Three of *Being and Event*, Badiou explains that inconsistent multiplicity describes a set that "is 'too large' to be counted as a set in the same way as the others. "Too large" is the metaphor of an excess of being-multiple over the very language from which it was to be inferred" (*BE* 41). Badiou surmises that the concept of the "too large" is Cantor's way of marking his work with an "ontological thesis," in which inconsistency "orientates thought toward the Infinite as supreme-being, or absolute" (*BE* 42). Cantor's solution, in the face of a restrictive and restricting paradox of construction, is to turn to the nature of multiplicity itself; the distinction of inconsistent and consistent multiples makes way for a pronouncement of being qua being as Absolute.

This ontological thesis, Badiou points out, can be interpreted along two lines. First, it may be the case that Cantor is viewed as "theologian," if his work "ties the absoluteness of being *not* to the (consistent) presentation of the multiple, but to the transcendence through which a divine infinity in-consists, as one, gathering together and numbering any multiple whatsoever" (*BE* 42). In this description, Badiou suggests that the proper place of the absolute is with consistency, the means by which the "there is" is guaranteed. However, the view of Cantor as theologian transforms the inconsistent multiple into an object of adoration; this position—the onto-theological position—is one Badiou is wholly opposed to.

A second interpretation of the ontological thesis assigns to Cantor a "brilliant anticipation" in which Cantor "saw that the absolute point of being of the multiple is not its consistency—thus its dependence upon a procedure of the count-as-one—but its inconsistency, a multiple-deployment that no unity gathers together" (*BE* 42). This reading of Cantor's work aligns being with inconsistent multiplicity and its science with consistency. In other words, the alternative to an onto-theology is a mathematical ontology, the commitments of which include the retrospective revelation of inconsistency through consistency, multiple by way of the count-as-one. Contrary to the absolute presence Cantor posits at the 'end' of his theorizing over the infinite, Badiou claims that the inconsistent multiple is marked by *no-thing*. This thesis, which is adjacent to that of the void set, is articulated in Meditations 4 and 5 of *Being and Event*.

Badiou's assessment of nothing and the void begins with a claim regarding "any situation in particular" (*BE* 52). Situations, as structured presentations, are such that inconsistency is obscured by the count-procedure. Beginning from the situation yields a false conclusion regarding being qua being: "Any situation, seized in its immanence, thus reverses the inaugural axiom of our entire procedure. It states that the one is and that the pure multiple—inconsistency—is not" (*BE* 52). In other words, if the ontological question is posed at the level of a situation, Badiou says, there would be no evidence of inconsistency at any point—the situation is such that is secured against the disruption inconsistency may generate.

To take these claims of the "immanence of the situation" further Badiou says, "inside what a situation establishes as a form of knowledge, being is being in the possibility of the one" (*BE* 53). This claim reinforces the strict *closure* the situation provides: what the situation "knows" (the content of the situation, the world "inside" the situation itself) is only its own terms, accounted for according to the reinforcing relations of belonging. A crude way of putting it is that *what is not counted* (that is, what does not appear in the situation) *does not exist*. This presents a significant difficulty, however, because if there are no means of exposing inconsistency, the count merely reinforces itself. That is, there is no evidence of the multiple from which the count-procedure emerges. Without some glimpse of the multiple, situations are mere tautologies and, perhaps worse (for Badiou), the One is reinforced. This is precisely the end Badiou's account aims to avoid. However, he probes the difficulty further, saying:

> To be sure, there is no antecedence of the multiple which would give rise to presentation because the latter is always already-structured such that there is only oneness or consistent multiples. But this 'there is' leaves a remainder: the law in which it is deployed is discernable as operation. And although there is never anything other – in a situation – *than* the result (everything, in the situation, is counted), what thereby results marks out, before the operation, a must-be-counted. It is the latter which causes the structured presentation to waver toward the phantom of inconsistency. (*BE* 53)

For Badiou, there is no "power" or "movement" (for lack of better terms) that would create a presentation. This reinforces the above point regarding the immanence of the situation and its nature as consistent and 'oned.' This second sentence indicates, however, a *relation* (again, for lack of a better term) between the situation and the multiple from which it is extracted—the situation is the result of an operation on this multiple. Badiou again reaffirms the nature of the situation as secured, but here he points out that the *operation* reveals something prior to the situation, which he describes here as "a must-be-counted." Put another way, the operation must have had something on which to operate to arrive at a structured result. Badiou's *fantôme* here, the specter that haunts the situation, is the inconsistent multiple, its spectral status confirmed by its 'disappearance' following the operation. As such, he explains that the inconsistent multiple must "be *nothing*" (*BE* 53), a site where existence is prohibited.

Badiou uses the status of inconsistent multiple as nothing to affirm the true thesis that "there is one-ness," and to discredit the "(false) thesis of the ontologies of presence, 'the one is'" (*BE* 53). In this way, the further distinction of the inconsistent multiple as 'nothing' confirms the trajectory of Badiou's ontological project against those positions he brackets, those positions seating the One as being qua being. There is also a finer distinction at play, however: "between the (true) thesis 'inconsistency is nothing,' and the (false) structuralist or legalist thesis 'inconsistency is not'" (*BE* 54). The false claim relegates inconsistency to the status of non-being (and the affirmation, following ontologies of presence, that the One alone is), while the true thesis simply asserts that the inconsistent multiple is "not *a* being" (*BE* 23).

The nothing, it turns out, is no *place*. Badiou explains, "it would be pointless to set off in search of the nothing" (*BE* 54). It will not emerge as a term in the situation, since that would require its having been subject to the count; the nothing is itself invisible in the situation. When Badiou claims that "the only thing we can affirm" regarding the relation of the situation and nothing is that "every situation implies the nothing of its all" (*BE* 54). The nothing trails the situation in the form of necessity, as a 'that-from-which' the situation is constructed; while the nothing cannot be discovered in the situation, nothing is necessary for a situation to be at all.

It is this relationship, between the inconsistent multiple and the situation, that allows Badiou to introduce the concept of the *void*, which he defines accordingly: "I term *void* of a situation this suture to its being. Moreover, I state that every structured presentation unpresents 'its' void, in the mode of this non-one which is merely the subtractive face of the count" (*BE* 55). In other words, the void of a situation is the *reminder* of the situation's relation to the nothing. Taking the void as a reminder has not only a reminiscent effect, but an orienting one as well—it is the *side* or *facet* of this presentation that indicates its disposition toward the inconsistent multiple.[8] As such, it is the "name of being—of inconsistency—according to a situation" (*BE* 56). By identifying the void, it is possible to now speak about inconsistency in the language appropriate to a mathematical ontology; this, in part, is the work of this text's Chapter 4.

Summary

In the preceding, I've discussed the value of inconsistent multiplicity for both Cantor and Badiou. These thinkers share an affirmation that ordinality must remain open and thus inconsistent; if it will be possible to generate large cardinals (the series of alephs) or transfinite ordinals (the series beginning with ω), a founding separation of closed, consistent multiplicity from open, inconsistent multiplicity is necessary. However, and as I explained above, Cantor's insistence on an Absolute or presence capable of closing inconsistency marks the condition at which Badiou's reading breaks away. The implication of open ordinality for Badiou is, straightforwardly, that the inconsistent multiple cannot be closed—no trap door exists in the articulation of inconsistency for transcendence to emerge. As an alternative, Badiou proposes *nothing* as the character of inconsistent multiplicity. As nothing, it resists closure or the 'internal' inscription of closed sets. Instead, the procedure of making-consistent, or counting, is facilitated by the presence of a void, or the "suture" to inconsistency present in any situation or structured presentation. There are two positive consequences that follow from Badiou's assertion of inconsistency as nothing; one consequence is *prospective*, the other *retrospective*.

The prospective consequence concerns assertions Badiou makes in Meditations 12–14 of *Being and Event* regarding the relationship of infinity to normal multiples, or a 'maximum' expression of consistency. The analysis of limit ordinals (ω_0, or the first transfinite ordinal) requires a persisting openness secured by an equally open inconsistency. The limit ordinal organizes an infinite series of successor ordinals, but it does not invoke a principle of closure. Rather, the limit ordinal *indexes* a border that allows the enumeration of ordinals to proceed. The presence of a limit ordinal, the assertion that infinite succession is possible from the border it creates, and its articulation in the context of the normal multiple, each pave the way for the subsequent meditations concerning singular multiplicity. In other words, the discussion of the limit ordinal is the fullest expression of a certain kind of consistent multiplicity, and forms a background against which the singular multiple and its issue of an evental site can be understood.

The retrospective consequence is a conformity of this contemporary ontological structure with an earlier political and theoretical structure Badiou/Peyrol asserts in "The Fascism of the Potato." In Chapter 1, I explained Peyrol's insistence that for a mass revolt to be successful, there must be a motivating element 'cut' from it. This framework transfers to the ontological division of in/consistency Badiou develops in *Being and Event* and beyond, and as such marks a 'periodization' in Badiou's own theorizing; mass is reconceived as inconsistent multiplicity, and class—that element creating a temporary and contingent unity from the mass—is reconceived as consistent multiplicity. The division of consistent from inconsistent multiplicity, the ascription of inconsistency to being qua being, and the assertion of consistency as an indication of presence (i.e., "there is one-ness") demonstrate a fidelity to the structure characterizing the explicitly materialist commitments of Badiou's earlier work. I say more about this connection in Chapter 4.

Conclusion

This chapter is the first of two 'explicative' chapters, in which I attend to matters of multiplicity in their mathematical and philosophical contexts. My focus here has been to demonstrate the way multiplicity serves to *structure* Deleuze's, Deleuze-Guattari's, and Badiou's respective accounts of being. In my analysis of multiplicity across Deleuze's oeuvre, I showed that whatever *is* is, at least in part, a multiplicity. By using Riemann's concepts of continuous, multiply-extended manifolds, I explained that these principles form the underlying shape and space of that which emerges. The virtual idea is built from connections that, when linked, form the conditions by which any empirical or conceptual instance is generated. An analysis of smooth space—whether in explicitly Riemannian terms or in terms of models expressing these commitments—shows that an understanding of space requires both its coordinates and the pathways connecting those coordinates. The trick in each of these cases, it seems, is to be able to search 'below' the empirical instance or discrete expression to see the network of connections that produce it. This shift is made possible in part by the transformation in spatial thinking Riemann's work occasions.

A similar commitment holds on Badiou's part, insofar as whatever *is* is also a multiplicity, but a *consistent* multiplicity, subject to the rules governing contradiction. It is posited alongside an inconsistent multiplicity, a multiplicity that remains open and—on Badiou's analysis—resists closure into an ultimate totality. This "innumerable immensity" is drawn from innovative thinking by Georg Cantor, who responds to a set-of-all-set paradox with a rewriting of the relationship of a set to that from which is created. The articulation of inconsistent and consistent multiplicities is necessary for an account of singular multiplicity and the evental site, the place of transformation and novelty in Badiou's system.

In the preceding, I have offered positive presentations of each program of multiplicity, designed to show the way multiplicity serves to structure accounts of being and a subsequent ontology. This presentation is warranted, given that the critiques I discussed in Chapter 2 often proceeded from incomplete or polemically-convenient presentations of an opponent's concept of multiplicity by their interlocutor. My focus in the next chapter takes up a similarly neutral presentation of procedure in their respective ontologies, beginning with Badiou's articulation of the void set and the axiomatic system that counts it.

Notes

1. Ferreirós (2007, 51) says Riemann's manner of presentation follows a "formal trend" in nineteenth century logic, insofar as Riemann's work depends on the position that "there is no reference to classes of objects

(individuals), only to classes of further concepts." The "formal trend" was intended as a restrictive application of logic to the procedural or formal use we are presently familiar with; it had, according to certain logicians of the day, been diluted by its use in Psychology and Epistemology. Per Ferreirós (2007, 48), this move constituted a "return to the Aristotelian conception and doctrines of logic." Ferreirós (2007, 52) points out that this mode of relationship "was absolutely common" in nineteenth century Germany in particular, and leads to its precise expression in Frege's comprehension principle. Arkady Plotnitsky (2009, 118) also identifies this reading of Riemann's initial definition according to the class-concept pairing, although he does not provide a history of this relationship. Ferreirós (2007, 41) identifies *Categories* 4b20 and *Metaphysics* 1077b18-20 as the Aristotelian sources of the distinction of continuous and discrete quantities; in the former, Aristotle discusses *poson* (quantity); in the latter, magnitude (*megethos*) and numbers (*arithmos*) are the concerns of mathematical propositions.

2. In addition to Deleuze's discussion of these ideas (*DR* 184–186), see also Daniela Voss (2013, 196–198).
3. Albert Lautman, *Les schémas de structure* (Paris: Hermann, 1938), 23, 34–35. The first quotation I discuss is from page 23, the second is found on pages 34–35 of Lautman's text. The full reference is found in *TP* 573.
4. My description is aided by Eric Weisstein's insights concerning the paradox in its attendant entry at Wolfram Mathworld.
5. Readers of Cantor's letter to Dedekind note that the order of presentation goes (1) *inconsistent multiplicity*, (2) *consistent multiplicity*, and (3) *well-orderedness*, with some intervening discussion about simply-ordered multiplicities and conditions for similarity.
6. Cantor later addresses the condition of simple order for three distinct numbers, saying, "of two distinct numbers α and β, one is always the smaller, the other greater, and that, if for three numbers we have $\alpha < \beta$ and $\beta < \gamma$, we also have $\alpha < \gamma$" (115).
7. Livingston points his reader to Michael Hallett's *Cantorian Set Theory and Limitation of Size* (Oxford: Oxford University Press, 1986). Hallett's discussion of Cantor's absolute includes Cantor's letter of 1887–1888 in which he claims a distinction between "increasable actual-infinite or *transfinite*" and "unincreasable actual-infinite or *Absolute*" (quoted in Hallett, 41). The latter magnitude cannot be apprehended, or is not 'rationally subjugable' (Hallett, 41) by human cognition or operation.
8. Badiou's readers know that it is possible for the void to appear—or "become localizable at the level of presentation"—in a situation. This possibility makes way for the event, an "ultra-one of a hazard" (*BE* 56).

References

Aristotle. 1984. *Categories*. Translated by J.L. Ackrill. In *The Complete Works of Aristotle: The Revised Oxford Translation*, vol. I, edited by Jonathan Barnes, 3–24. Princeton, NJ: Princeton University Press.

Badiou, Alain. 2005. *Being and Event*. Translated by Oliver Feltham. London: Continuum. Originally *L'être et l'événement*. Paris: Éditions du Seuil, 1988. (Cited in text as *BE*)

———. 2006. *Briefings on Existence: A Short Treatise on Transitory Ontology*. Translated, edited, and with an Introduction by Norman Madarasz. In *Intersections: Philosophy and Critical Theory Series*, edited by Rodolphe Gasché. Albany, NY: State University of New York Press. Originally *Court Traité d'ontologie transitoire*. Paris: Éditions du Seuil, 1998. (Cited in text as *CT*)

Biagioli, Francesca. 2016. *Space, Number, and Geometry from Helmholtz to Cassirer*. Archimedes 46: New Studies in the History and Philosophy of Science and Technology. Switzerland: Springer International Publishing.

Burali-Forti, Cesare. 1977. "On Well-Ordered Classes (1897a)." Translated by Jean van Heijenoort. In *From Frege to Gödel: A Source Book in Mathematical Logic*, edited by Jean van Heijenoort, 111–112. Cambridge, MA: Harvard University Press.

Cantor, Georg. 1977. "Letter to Dedekind (1899)." Translated by Stefan Bauer-Mengleberg and Jan van Heijenoort. In *From Frege to Gödel: A Source Book in Mathematical Logic*, edited by Jean van Heijenoort, 113–118. Cambridge, MA: Harvard University Press.

Deleuze, Gilles. 1994. *Difference and Repetition*. Translated by Paul Patton. New York: Columbia University Press. Originally *Différence et repetition*. Paris: Presses Universitaires de France, 1968. (Cited in text as *DR*)

Deleuze, Gilles, and Félix Guattari. 1981. *Rhizome: An Introduction*. Translated by Paul Patton. *I & C* 8 (Spring): 49–71. Originally *Rhizome: Introduction*. Paris: Les Éditions de Minuit, 1976. (Cited in text as *R*)

———. 1987. *A Thousand Plateaus*. Translated by Brian Massumi. Minneapolis, MN: University of Minnesota Press. Originally *Mille Plateaux*. Paris: Les Éditions de Minuit, 1980. (Cited in text as *TP*)

Ferreirós, Jose. 2007. *Labyrinth of Thought: A History of Set Theory and Its Role in Modern Mathematics*. Basel: Birkhäuser Basel.

Gillespie, Sam. 2008. *The Mathematics of Novelty: Badiou's Minimalist Metaphysics*. Anamnesis. Melbourne: re.press.

Livingston, Paul. 2012. *The Politics of Logic*. London: Routledge.

Merzbach, Uta, and Carl Boyer. 2011. *A History of Mathematics*, 3rd ed. Hoboken, NJ: Wiley.

Plotnitsky, Arkady. 2006. "Manifolds: On the Concept of Space in Riemann and Deleuze." In *Virtual Mathematics: The Logic of Difference*, edited by Simon Duffy, 187–208. Edinburgh: Edinburgh University Press.

———. 2009. "Bernhard Riemann's Conceptual Mathematics." *Configurations* 17 (1–2): 105–130. https://doi.org/10.1353/con.0.0068. Accessed 12 June 2013.

Riemann, Bernhard. 1873. "On the Hypotheses Which Lie at the Bases of Geometry." Translated by W.K. Clifford. *Nature* VIII (183–184): 14–17, 36–37.

Russell, Bertrand. 1977. "Letter to Frege (1902)." In *From Frege to Gödel: A Sourcebook in Mathematical Logic*, edited by Jean van Heijenoort, 124–125. Cambridge, MA: Harvard University Press.

Scholz, Erhard. 1992. "Riemann's Vision of a New Approach to Geometry." In *1830–1930: A Century of Geometry: Epistemology, History, and Mathematics*, 22–34. Berlin, Heidelberg: Springer.

Van Heijenoort, Jean. 1977. Introduction to "Georg Cantor: Letter to Dedekind (1899)." In *From Frege to Gödel: A Source Book in Mathematical Logic*, edited by Jean van Heijenoort, 113. Cambridge, MA: Harvard University Press.

Voss, Daniela. 2013. *Conditions of Thought: Deleuze and the Transcendental Idea.* Edinburgh: Edinburgh University Press.

Zermelo, Ernst. 1977. "Investigations in the Foundations of Set Theory (1908a)." Translated by Stefan Bauer-Mengelberg. In *From Frege to Gödel: a Sourcebook in Mathematical Logic*, edited by Jan Van Heijenoort, 199–215. Cambridge, Mass: Harvard University Press.

CHAPTER 4

Procedures: One, Multiple, Subtraction

In Chapter 3, I explained Deleuze's and Badiou's respective theories of multiplicity. Though it is widely recognized that they differ over what constitutes multiplicity, my interest was to link their preferred versions of multiplicity to ontological *structure*. The present chapter is focused on the procedure prescribed by structure, or the mechanism(s) Badiou and Deleuze use to account for *one* in relationship to the *multiplicity* forming the structure of ontology. I begin where I left off in the previous chapter by examining the tradition of one-multiple relations in Badiou's oeuvre. Badiou's affirmation of a one that results from counting the multiple, or *one-ness*, is a necessary step toward transformation. An event requires an evental site, which requires at base a situation, or an organized and counted multiple. I focus on the system Badiou develops in *Being and Event*, where the work of defining consistent multiplicity with the count-procedure requires a point at which inconsistent multiplicity is exposed to operation; this is the void set and its mark Ø, necessary for ontological structure. This process is aided by the axioms of Zermelo–Fraenkel set theory. Though it is prohibited from being qua being, Badiou affirms one-ness as a function in mapping any event and describing the pathway from being to transformation.

By contrast, Deleuze and Deleuze's writing with Guattari accommodate any one insofar as it is held in suspension with the multiplicity occasioning it. Deleuze and Guattari maintain this commitment, for example, in their analysis of the war machine in "Rhizome," or the series of examples—including mathematical ones—in Plateau

B. Vartabedian, *Multiplicity and Ontology in Deleuze and Badiou*,
https://doi.org/10.1007/978-3-319-76837-3_4

"1440: The Smooth and the Striated" in *A Thousand Plateaus*. In his earlier work, Deleuze's commitments to division and overturned Platonism in *Difference and Repetition* develop the concerns on which the later texts build. In these cases, whether early or late, Deleuze and Guattari insist that any one is an *expression* of the multiplicity to which it is tethered. To identify the multiplicity generating an expressed one, they prescribe a procedure of subtraction, or $n - 1$, to "make the multiple." This injunction is familiar from the reading of "Rhizome" I offered in Chapter 2. However, Deleuze and Guattari affirm *consolidations* of multiplicity insofar as these tethers remain intact; I illustrate the relationship of consolidation and subtraction in the cogito diagram illustrating "Example 1" in *What is Philosophy?*

Badiou's One-ness: Void, Scission

In the preceding chapter, I discussed the Cantorian distinction of the inconsistent multiple from consistent multiples and explained their function in Badiou's ontological project. Badiou assigns inconsistent multiplicity to being qua being, while consistent multiplicity is the ontological position at which existence is established. I identified the origin of this distinction in Georg Cantor's novel solution to the Burali-Forti paradox; consistent multiples are those that can be closed or made into sets without resulting in a contradiction, while an inconsistent multiple cannot be closed without contradiction and thus remains open and infinite. I discussed Zermelo's axiom of separation, which dictates the relationship between inconsistent and consistent multiples such that consistent multiples are separated from an already-available multiple. For Badiou's purposes, Zermelo's work authorizes a condition under which set-theoretic ontology—the work of constructing situations—can proceed. Finally, I explained the significance of *no-thing* in relation to inconsistent multiplicity and with respect to Badiou's overall project; it is in the *no-thing* that Badiou breaks with Cantor (on the one hand) and introduces the void as a foundational piece of his ontological program (on the other).

However, and though the resources for understanding the structure of ontology in Badiou's work are rich, it is insufficient to claim that the identification of two types of multiplicity describes a procedure for relating one and multiple in Badiou's work. Indeed, this distinction only identifies the ground over which determination of any one is possible. Badiou's demonstration of the void and its mark Ø identifies the procedural component in Badiou's ontology. As the "first existent" in Badiou's system, it functions as the object of an initial count-procedure and persists irrespective of the complexity that may append to that structure. With the more contemporary accounting for the count in Badiou's work established, I return to his earlier preoccupation with the work of *scission*, or his analysis of the dialectic by division to show the ways Badiou's thinking on the relation of one and multiple marks an evolution in the operation of dialectical thinking across his oeuvre.

From Multiple to One: Void

Badiou's work maintains a consistent commitment to a theoretical configuration affirming *multiple* or *mass* and some temporary one emerging from this multiple. In the early work, from the 1965s "The (Re) Commencement of Dialectical Materialism" through the lectures collected in *Theory of the Subject*, this affirmation is primarily political and motivated by Badiou's explicitly Marxist and Maoist commitments. By 1985, however, Badiou seems to have abandoned these explicit commitments. Indeed, and as Oliver Feltham (2008, 85) points out, the appearance of the short text *Peut-on penser la politique?* "seals the almost complete disappearance of Marxist vocabulary—party, dialectic, revolution, proletariat—from Badiou's work." Badiou affirms a necessary transformation of dialectical thinking, given the turn in the Soviet Union and associated Bloc countries toward parliamentary politics. This new version of the dialectic, Badiou (1985, 86) explains, "will be recognized first of all by its conflict with representation. Such thinking tracks down the unrepresentable point in its field, from which it turns out that one touches on the real" (translation mine). Instead of seeking an internal condition to promote transformation (e.g., the class condition discussed in "The Fascism of the Potato"), Badiou's interest shifts toward finding an unaccounted-for point as this source.

This short description in *Peut-on penser la politique?* bears a purposive similarity to the relation Badiou prescribes for science and ideology in his earlier work. For example, in "The (Re)Commencement of Dialectical Materialism," (1965), Badiou situates science and ideology in an oscillating relationship, one in which the transformations science brings to ideology are eventually absorbed and thus require new scientific transformations to re-arrange the order. Badiou explains, "If science is a process of *transformation*, ideology—insofar as the unconscious comes to constitute itself therein—is a process of *repetition*" ([1965] 2012, 147). This play of transformation and repetition requires positive intervention. Twenty years later, however, Badiou's revision to dialectical thinking reveals a process committed to seeking the point of transformation something like ideology or social-political discourse has covered over. By the time of *Peut-on penser la politique?*, Badiou insists that thought must search for a point responsible for disruption and transformation; the presentation in 1985 is expressed in language consistent with Badiou's repositioning theory toward the "pure multiple," work that is first and foremost ontological and inaugurated in *Being and Event*.

In Part IV of *Being and Event's* Introduction, Badiou accounts for the way this search for the unpresentable point proceeds. The "logicist" commitments that occupied his earlier work, Badiou explains, were inconsistent "with the clear Lacanian doctrine according to which the real is the impasse of formalization. I had mistaken the route" (*BE* 5). Badiou's concern, at least from this vantage, is to discover the *limits* of thought, the point at which formal structures no longer apply. No such limit is identified in Badiou's earlier work; there, dialectical thinking is concerned with formal scientific structures that directly act on static ideology in order to transform it (Brassier 2005). However, in *Peut-on penser la politique?*, thought's task is to pass through structures of representation to identify the point of access to the real.

Badiou's transformation also registers a shift in his assessment of mass or multiple. Already convinced of the efficacy of axiomatic thinking—a commitment as early as *The Concept of Model* (1968)—Badiou identifies multiple as the background of his approach. He elaborates on this in *Being and Event*, saying,

> I came to the conclusion that the sole manner in which intelligible figures could be found within was if one first accepted that the Multiple, for mathematics, was not a (formal) concept, transparent and constructed, but a real whose internal gap, and impasse, were deployed by the theory. (*BE* 5)

The multiple, as Badiou describes it here, must itself remain impervious to structure except for its "internal gap and impasse," or the point at which the multiple is exposed as such. This is the gap, and its impasse is this same point understood as a limit of thought. Though these are features of the multiple, Badiou closes this claim by insisting that the gap and impasse are legible in the theory of the multiple. This leads Badiou to the void set and its proper name Ø. The void is 'that from which' consistent multiples are constructed, and it also serves as the boundary for formal thinking. As such, the void serves to bridge inconsistent and consistent multiples; however, and before turning to Badiou's explicit account of the void, it is perhaps helpful to see what it is not.[1]

Traditionally, and when it is aligned with ancient atomists and Lucretius, void is an empty field through which bodies circulate. According to Aristotle, atomists like Democritus and Leucippus maintained "the full and the empty are the elements, calling the one being and the other non-being" (1984, 985b5-6). The element associated with being is full and solid and consists in a body; the void, the element associated with non-being, is an empty field in which atoms are organized. In *De Rerum Natura*, Lucretius explains that the universe "as it is in itself, is made up of two things; for there are bodies, and there is void in which bodies are and through which they move this way and that" (1975, 418–421). For these early thinkers, the void is an absence or emptiness, though an emptiness is necessary as the bodies that move in it. Of Democritus and Leucippus Aristotle explains, "they say that what *is* is no more than what is not, because body no more is than the void" (1984, 985b8-9). Though the principle of existence is attributed to bodies, the emptiness of void is required to trace movement, integration (or disintegration), and change. Lucretius and the atomists depart from Parmenidean strictures concerning being, since they each maintain the reality of *both* being and non-being and thus affirm the possibilities for these to be objects of thought; in the Parmenidean account the goddess has barred human thought from even considering non-being.

Like the atomists and Lucretius before him, Badiou affirms the void as both an object of thought and a necessary component of any ontology. Unlike these precedent thinkers, however, when void is deployed in an ontological context it bears no traces of a material connection, and though Badiou refers to it as a "multiple of nothing," and marks it Ø, it is not an instance of non-being. Though Badiou maintains what I have called above a "dispositional affinity" with these earlier positions, the specific content of the void differs from these precedent presentations. I now turn to Badiou's demonstration of this idea and discuss its significance as the procedure on which Badiou's ontological turn is founded.

On the Mark Ø

In the previous chapter, I explained the conceptual difference between inconsistent and consistent multiples and discussed the axiom of separation as a principle for understanding the status of a consistent multiple vis-à-vis an already-available multiple. The axiom of separation distinguishes "that 'on the basis of which' there is presentation," that is the inconsistent multiple, from "the result of presentation," or consistent multiple(s). However, these concepts are static; in the matter of actually constructing a consistent multiple, a fourth component is required. Badiou introduces this as "an unpresentable yet necessary figure which designates the gap between the result-one of presentation and that 'on the basis of which' there is presentation" (*BE* 55). In other words, the fourth component is the *mark* of separation between inconsistent and consistent multiples. Badiou then names this gap "the nothing particular to the situation, the unlocalizable void point" that clarifies "the situation is sutured to being" (*BE* 55). In other words, the void—this fourth component required for the construction of consistent multiplicity—maintains a "present absence" that identifies the situation as multiple and as such as separated from inconsistent multiplicity.

Badiou posits the void, first, as a "pure proper name" (*BE* 59). As a proper name, this designation is such that it refers only to itself; the content of this reference, however, is the no-thing of the inconsistent multiple. The void as pure proper name will thus mark each situation with no-thing, accomplishing the "suture" of any situation to being as prescribed in Badiou's description. However, and in spite of this appellation, it's necessary to note two things: first, identification of a pure proper name serves only as a link to inconsistent multiplicity, and is insufficient to establish existence—more work is required on this score; second, the void will not serve as a pathway to being qua being, but rather a procedural stop, marked Ø.

The mark Ø is a symbolic intervention attributed to the Bourbaki group, which originated at the École normale supériéure in the late 1930s. Bourbaki member André Weil claimed responsibility for the selection of Ø to designate the empty set, a move part of Bourbaki's larger efforts to transform set-theoretic language toward more precision.[2] Such a move is in-line with the broader moves of formalism, particularly those developed by and following Frege in the wake of Cantor's problematic inexactness. Badiou has long admired the work of the Bourbaki group, noting that their intervention at the junction of arithmetic and geometry inspires the broader commitments of his ontological project. He explains that their work

> … makes a distinction from the outset between algebraic structures, which are the possible structures (addition, subtraction, division, root extraction, etc.) that enable calculations, and topological structures, which make it possible to think spatial arrangements (neighborhoods, inside and outside, connections, the open and the closed, etc. (2016, 48)

Readers of Badiou's oeuvre see the links of this recent *paean* to Bourbaki as valuable for mapping Badiou's own project; like this intrepid group of mathematicians, Badiou's own projects depend on the respective trajectories engaging algebra and the "possible structures" of calculation (in *Being and Event*) and those of topologies (in *Logics of Worlds*). However, Badiou's deployment of Bourbaki is not only honorific—it also has procedural significance for the ontological project.

The empty set Ø is the complement of any set E (Bourbaki 1968, 349); intuitively, and given that a set is an ensemble of elements and as such is 'full,' it is reasonable to see the empty set in a negative relation to this set E. Procedurally, the complement is identified in Bourbaki by subtraction and defined with subsets (1968, 72). Take, for example, the set "Becky's nieces and nephews," containing the elements (B, N, T, F, W). Identifying the subset "Becky's nieces" draws from the original set elements (N, T, W). The complement is thus the subset of elements to which the identified property *does not apply*; the subset (B, F), or "Becky's nephews" is the remainder, and the complement of "Becky's nieces." At the level of set Bourbaki (1968, 349) explains, "some properties, for example $x = x$, are true for *all* elements of E." Here, Bourbaki identifies a property that holds for the entirety of elements in the set E. Since the work of determining a complement required subtraction from the given set all elements and subsets with the identified property, the complement of set E (the set for which the property $x = x$ is true for all elements) is the set for which the property $x \neq x$ is true. The complement is, therefore, a set with no elements.

For Badiou's work, two aspects of this operation are of value. First, the operation by which a complement is determined is subtraction; it is the result of subtracting a subset from the larger set. In the case of determining the complement at the level of the set, the set E, when subtracted from itself, yields the set with no elements, Ø. Second, it is worth noting (especially in relation to Badiou's claims concerning the presence of Ø in any situation) that complements determined subtractively must be already *available* to serve as complement. Bourbaki's analysis shows the empty set Ø present as complement, a claim consonant with the requirements Badiou prescribes concerning the void and its mark Ø as a foundational component in any consistent multiple.

In an interview with Tzuchien Tho, Badiou (2007, 99) explains, "the void is the point on which we found the constructible sets which allow us to unfold the characteristics of pure being." In terms I've been using above, the void is a site from which consistent multiplicity is built. However, the void's presence in a consistent multiple is an element antithetical to the situation in which it appears; in other words, the void maintains the link to inconsistent multiplicity. The mark Ø is the point of no-thing that must be carried through any situation; it is that tether Badiou sought in order to relate presentation to operation. Badiou cannot simply assert, however, that Ø exists. Rather, it must be found via the "principal Ideas of the multiple," or in axioms (*BE* 60).

A Multiple of Nothing and Its Axiomatic Expression

The conceptual commitments to inconsistent and consistent multiples are constraints on the way Badiou defines the void. Though he asserts its status as a "pure proper name" and deploys the mark Ø, the void cannot be issued by inconsistent multiplicity as a one without having been counted; this contradicts the fundamental commitment to one as a *result* of ontological operation and jeopardizes the nature of inconsistent multiplicity itself. Instead, Badiou specifies that the void "auto-declares itself in the form of the multiple, despite there being *nothing* numbered by it" (*BE* 59). The discussion of the mark Ø in connection with the empty set above clarifies the last part of this claim; Ø is the empty complement of any set whatever, and it is precisely this emptiness, this *no-thing*-ness that inscribes the boundary for inconsistent multiplicity on the one hand. On the other, the void's "auto-declaration" occurs in a mark legible to mathematical structuring. It is in this valence, that of the language of the multiple, that the void is apparent as a "'multiple' of nothing" (*BE* 67).

The empty set has no differentiating element; nothing belongs to it and nothing *can* belong to it. As Badiou points out, "the structure of the statement that inscribes the 'first' existence is thus, in truth, the negation of any existence according to belonging" (*BE* 67). Put another way, the relations that maintain ordinary sets must be negated to find the existent on which these sets are founded. In light of this, it is reasonable to wonder how existence follows from negation. The statement Badiou presents is an axiom, which follows the tradition of Zermelo's axiom of the null set and Bourbaki's axiom of the empty set. The axiom, in natural language, reads, "there exists that to which no existence can be said to belong" (*BE* 68); presented formally, the statement is $(\exists\beta)\ [\sim(\exists\alpha)(\alpha \in \beta)]$, there exists a set β for which no element α that would belong to β exists.[3] This formal presentation explicitly connects the procedure to an axiom system since, as Badiou notes, the void set "will begin with an existential quantifier (thereby declaring that being invests the Ideas)" (*BE* 68). On account of the formal presentation and its axiomatic status, the void set is effectively 'separated' from the inconsistent multiple to count, though it retains is peculiar place as *link* or *trace* of any ontological construction to being qua being, or the inconsistent multiple.

Badiou's axiomatic presentation of the existence of the void set is the first and foundational step in linking one and multiple; it is now inscribed in the language of ontology, both in terms of content (the appearance of a mathematically legible symbol, Ø) and form (i.e., the use of an axiom). The procedural transformation toward consistency—that is, the work of moving this from a structuring concept to a content-filled object of ontological inquiry—is a movement "from being-without-one" to the one of a structured situation, governed by the relations of belonging and inclusion.

From Nothing to Number

My present concern is to discuss the movement to consistent multiples and subsequently to number, the results of the process linking multiple to one in Badiou's ontology; in other words, I will show how Badiou moves from a "'multiple' of nothing" (*BE* 67) to *one*. This requires attention to the supplementary relation of inclusion, the axiom of extensionality, and the axiom of the power set, as these appear in *Being and Event*.

Before introducing or discussing these tools, Badiou reminds his reader that belonging is "the originary relation" (*BE* 81). Belonging is a strong relation, insofar as it defines the whole of a set. The elements *B*, *N*, *T*, *F*, and *W* each belong to and so define the set "Becky's Nieces and Nephews." Inclusion, according to Badiou, "indicates that a multiple is a submultiple of another multiple" (*BE* 81). If belonging describes the whole, inclusion describes the parts; the elements *B*, *N*, *T*, *F*, and *W* are each parts of the set "Becky's Nieces and Nephews." Though it builds from a multiple, the inclusion relation results in "a multiple essentially distinct" from the set it is a part of (*BE* 83). There are two axioms that support this claim. First is the axiom of extensionality (in Bourbaki, the axiom of extent [1968, 67]), which asserts that two sets are equal if and only if these contain identical elements (*BE* 60–61). Second is the axiom of the power set, which indicates that the subsets of any given set can themselves be counted as a set: {*B*, *N*, *T*, *F*, *W*} counted as individuals {*B*}, {*N*}, {*T*}, {*F*}, {*W*}; counted by parentage {*B*, *N*, *T*} and {*F*, *W*}; by gender {*B*, *F*} and {*N*, *T*, *W*}; by age {*B*, *N*} and {*T*, *F*, *W*}, and any other relevant parameters that apply to these elements. The combination of each of these *and* the original set constitutes the power set; as such, the eleven subsets I've drawn from the original are each included in the original set. The axiom of the powerset shows that the subsets organized from the original set are always more numerous than the multiple from which they are drawn. However, because they do not contain identical elements are distinct from one another (per the axiom of extensionality); put straightforwardly (and for example), the elements in the set {*B*, *N*} do not correspond exactly to the elements in the set {*B*, *N*, *T*, *F*, *W*}, and so the sets are not equal.

Badiou explains that the void set Ø is included in—and is thus a part of—any set (*BE* 86). My previous discussion of the empty set identified it as a *complement* of any set; Bourbaki shows that the empty set remains when any set is subtracted from itself. This finding is consistent with Zermelo's claim that "the null set is a subset of every set M" (1977, 202). To the eleven subsets associated with "Becky's Nieces and Nephews" I add a twelfth in Ø, a placeholder for the complement (in Bourbaki's analysis) and a register of the void (in Badiou's).

Badiou explains that inclusion reveals a second feature of the void set; Ø "possesses a subset, which is the void itself" (*BE* 86). Returning to the illustration of "Becky's Nieces and Nephews" above, I identified the subsets {*B*}, {*N*}, {*T*}, {*F*}, {*W*}; Badiou asserts that the same move is available for the void set. It has a subset, which Badiou writes *p*(Ø), and this subset indicates the presence of a *part* of Ø. This is perhaps a peculiar claim, augmented by Badiou's assertion that *p*(Ø) "solely announces that everything which is presented, including the proper name of the unpresentable, forms a subset of itself, the 'maximal' subset" (*BE* 88). These single subsets are maximal insofar as these express only the included element; in other words, the element both belongs *and* is included in the count. The subset {*T*}, for example, is only and totally {*T*}, not {*T*} in a larger group organized by property (e.g., gender, parentage, or age).

In the case of the subset *p*(Ø), the relationships are a bit trickier to track. The subset {*T*} is not empty, and its status as both element and part of the larger set is relatively uncontroversial. However, Ø has no content, nothing belongs to it, and it is identical to itself. The subset *p*(Ø) , though, is *not* empty. Its element is the proper name of the void, something it re-presents in its inscription; Badiou shifts to writing it {Ø}, and the axiom of extensionality demonstrates that Ø and {Ø} are distinct multiples. This describes the procedure of "forming-into-one of Ø" (*BE* 91). If the void set presents nothing, then the subset of the void set presents *one* 'thing,' which is the proper name of the void. This implication is repeated: if Ø, then {Ø}; if {Ø}, then {{Ø}}, and *numbers* are expressed by axiomatic operation on the void set. The move to consistency requires the void and axioms of separation, extensionality, void set, and power set. However, and before moving on from Badiou's ontological accounting, it is necessary to note that the counts {Ø} and {{Ø}} are each subject to a second count. "All situations," Badiou explains, "are structured twice. This also means: there is always both presentation and representation" (*BE* 94). By this requirement, {Ø} is represented as 1, {{Ø}} is represented as 2, and so on. The move from *nothing* to *number* is complete.

Up to this point my interest has been to show how one and multiple are related in Badiou's contemporary ontology. I have been especially interested in showing this relation in light of the shift Badiou prescribes for dialectical thinking in the 1985s *Peut-on penser la politique?* This prescription indicates that dialectical thinking is tasked with pursuing an 'unrepresentable point,' or the presence of the Real in any situation; in Badiou's ontology, this point is the void set, which acts as a mediating boundary between the infinite inconsistent multiplicity and consistent multiples, the closed sets at the foundation of Badiou's ontological project. When demonstrated alongside Zermelo's axioms of separation, extensionality, and powerset, the void set authorizes the construction of consistent multiples; the void set presents the *no-thing* of inconsistency to the axioms of set theory, thus bridging the gap between being qua being and its science. The void set also identifies the limit of ontology's reach—to the void set, but not beyond it—and as such identifies the impasse of thought when it comes to being qua being. To be sure, the search for this void is prescribed insofar as its irruption in a situation is the index of an event; but for the purposes of this study, I am interested in showing the void as a key element in Badiou's procedures for relating multiple and one.

This set of mathematical procedures, developed in concert with a reorientation of dialectical thinking, is not the only procedure for relating one and multiple in Badiou's *oeuvre*. In Chapter 2, I discussed Badiou's doctrine of *scission* in the critique Badiou—writing as Georges Peyrol—levies against Deleuze and Guattari. In what follows, I offer an account of scission as Badiou develops it in the 1975 lecture that opens *Theory of the Subject*. Though scission is primarily developed as the framework underwriting the emergence of a subject, I attend to its presentation of *one* from a *multiple* (in this case, a two). This account has two benefits for my present examination: first, scission prefigures a trajectory Badiou acknowledges and transforms in the more contemporary ontology; second, it clarifies the position from which Peyrol critiques Deleuze and Guattari's account of multiplicity.

From Multiple to One: Scission

The set of lectures that open *Theory of the Subject* are broadly concerned with the "Place of the Subjective," an identification of the procedures and components that contribute to the determination or existence of *something*. Badiou's inquiry depends on distinguishing two dialectical matrices "at the heart of the Hegelian dialectic" (*TS* 3). One, *alienation*, moves thought forward; the other, *scission*, moves thought down. As Badiou explains, in scission, "…there is no unity that is not split. There is not the least bit of return into itself, nor any connection between the final and the inaugural" (*TS* 4). Scission essentially freezes the movement of thought at a moment of temporary unification to attend to the components constituting that moment; in other words, scission requires a stoppage and analysis of the increasingly complex content of any unity at any stage of the dialectic. Badiou selects the interpretive matrix of scission to investigate "the notion of the 'something', which is the first form of being-there in Hegel's *Logic*" (*TS* 4).[4] The analysis of dialectical thinking requires, per Badiou, a presentation of *existence*.

In Hegel's *Science of Logic*, existence is the second move in unfolding the doctrine of being; the first is the movement from *being* to *nothing* to *becoming*. Hegel (2010, 59/21.69) asserts that because pure being is without determination, it "is pure indeterminateness and emptiness," and as such is nothing; nothing, though also empty and indeterminate, requires a thought to fix its status. The procedures of thinking about being and thinking about nothing are indicative of their sameness: "Nothing," Hegel says, "is therefore the same determination or rather absence of determination, and thus altogether the same as what pure *being* is" (2010, 59/21.69). Their sameness thus allows Hegel to identify the "immediate vanishing of one into the other" as *becoming* (2010, 60/21-69-70). By identifying this transformation of being into nothing (and vice versa), becoming discloses a new unity and the "truth" of being and nothing; the dialectical procession redefines being as *coming-to-be* and nothing as *ceasing-to-be.*

The dialectic of existence begins from becoming, coming-to-be and ceasing-to-be, represented as a unity labeled "determinate being." Hegel (2010, 83/21.97) notes that existence is "the simple oneness of being and nothing." Instead of analyzing becoming from the simple plurality of coming-to-be and ceasing-to-be, these are carried forward as an immediate unity. Hegel (2010, 83/21.97) identifies this unity as "*Dasein*; according to its [German] etymology, it is being (*Sein*) in a certain *place* (*da*)." As I will discuss below, the claim that existence is a matter of determination in place is significant for Badiou's 1975 analysis. For Hegel, however, the assertion of this determinate unity as *Dasein* is insufficient on its own for inaugurating existence; *Dasein* must be posited or expressed "in negation, in something and *other*" (Hegel 2010, 84/21.97). Put another way and in order to exist, *Dasein* must be both the place it is in *and* the place in which it is not. This qualitative distinction is significant for the account of existence in the *Logic*, insofar as the negation in here is determinate *non-being*. Hegel (2010, 85/21.99) defines this negation, saying it is "still a quality but one that counts as a lack and is further determined as limit, restriction." The simple assertion of *Dasein* is qualitatively transformed by that which it is not, determinate non-being; it is inscribed by an identification of its boundaries. The temporary unity of determinate being (*Dasein*) and determinate non-being is *something*, the inaugural moment of existence (Hegel 2010, 88/21.102).

My description of Hegel's work in the preceding two paragraphs depends on the interpretive matrix of alienation, in which thought is ushered forward by the identification of a contradiction (e.g., being is revealed, on analysis to be nothing). The matrix of scission pauses at the unity of *something* to interrogate the two moments constituting it. The inscription of existence as a unity of ***being-there*** and ***being-other-than-there*** allows Badiou to claim, "Everything that exists is thus at the same time itself and itself according to its place" (*TS* 8). I can re-interpret the moments of the dialectic identified above as follows: the conditions for the existence of anything are (a) the thing itself, the unified presence of coming-to-be and ceasing-to-be, which is qualitatively indistinct; and (b) the thing in place, "returned to itself" from its limit in non-being, a recognition of that which it lacks. This latter move qualitatively distinguishes the thing precisely on account of its placement. Badiou's reading of scission isolates these dialectical components to develop a formula, A and A_p, where A identifies a thing A-as-such, and A_p as A-in-another-place. The components A and A_p allow Badiou to register "the minimum primary differential" required for existence, which Badiou explains by saying, "***this*** only differs from ***that*** by the statement of difference, by the literal placement" (*TS* 6). Badiou signifies this as AA_p, which should be read as A, a thing as such, times A_p, a thing in another place.

Badiou devotes some attention to the notation A_p that he introduces to mark A-as-other. The place $_p$ can be expanded to infinity, though a quantitative reduplication (e.g., A_{p1}, A_{p2}, A_{p3}, ...A_{pn}) does not transform A_p beyond its status as other. This initial signification is not itself scission, but rather the marker of minimal difference so that determination might be accomplished. Per Badiou, "a constitutive scission," or the formula establishing existence, is $A = AA_p$ (*TS* 6), follows from the identification of these components and their repetition. The relation of contradiction is present in scission between A and "the space of placement P" (*TS* 7), pointing to the world in which A is placed; the qualitative limitation in Hegel's account of something, where something and something-other are the terms in play, is realized for Badiou in this determining background.

Badiou's analysis of scission sees the place or world operating as a third term that defines something-other (A_p) in particular. He supports this with a political example, describing place as "the bourgeois world, imperialist society," and situating this background as "the true contrary of the proletariat" (*TS* 7). In the bourgeois world, the proletariat (A) is reduplicated as the bourgeois (A_p); considered in reverse, the proletariat is "the principal productive force" and "antagonistic political pole" in the bourgeois world (*TS* 7). Scission articulates the relationship of something A, something-other A_p, and the world in which these interact

P as constitutive of an existence. It depends on an identification of difference holding between *A* and A_p, and the formula $A = AA_p$ expresses "the only form of existence of something in general" (*TS* 10). The formula $A = AA_p$ replaces the Aristotelian identity $A = A$. Something, for Badiou, is not identical to itself, but exists when it is seen alongside its limitation.

From this assertion, Badiou articulates a series of iterations for scission that describes a series of "great dialectical concepts" that form the foundation on which subjectivity emerges. These include ***determination***, which is scission "thinkable only from the indexed term … $A_p(AA_p)$"; a "determination of the new, $A_p(A)$," or *force*; and a *relapse* or reassertion of place: "$A_p(A_p) = P$" (*TS* 10). Badiou's interest here is to describe the set available relations between something, its limitation (as placed elsewhere), and the world in which this is placed. There is a normalizing or neutralizing function in determination's application of A_p, where the "something other" is doubled over something *A*; there is also an overriding function in the relapse—what Badiou will call the *splace* of determination—in which something *A* is prevented from appearing altogether. Badiou's focus is, perhaps unsurprisingly, an opportunity in which something *A* is neither neutralized nor overridden; if transformation is at stake, then the determination of the new—*force*—must go unrecognized or unassimilated by the world.

Badiou's alignment with Hegel on this score is suggestive of access to a 'retrospectively present' multiple. Nina Power (2006, 197) explains that scission is used "as the basis of an attempt to found a distinction between 'something' and 'another thing' … the dialectic is first and foremost a process, not of negation and the negation of negation, but of internal division." Though Badiou's alignment here is with the Maoist dictum that "One divides into two," the reading of the dialectic as *scission* and not *alienation* at least gestures to the possibility that scission can be applied at any point of the dialectic. Put another way, readers of Hegel know that each moment of the dialectic carries forward the previous and precedent moments, and so the unity accomplished at the end of the analysis of *something* is a unity that follows from the previous of moments the dialectic has passed through to arrive at that unity. Badiou's analysis of scission suggests that a unity of this sort—even if it is temporary—could be analyzed to identify the component parts.

Scission, Void, One-ness

My attention in the preceding section has turned on procedures for one-ing multiples, evident in Badiou's ontological project in *Being and Event* and gestured to his earlier political inquiry. Badiou's doctrine of scission, a hallmark of his political theorizing, prescribes a slowing or freezing of the movement of negation to identify a temporary unity and the multiple constituting it. Scission proposes that this relationship between one and multiple is more-or-less direct; the conditions for transformation and the condition for suppression are already present in the constitutive formula. Further, the various combinations for determination, force, and relapse are of the defined and available elements. The only additional requirement is proceeding from a matrix of scission rather than a matrix of alienation; alienation, which emphasizes movement by negation will never slow sufficiently to identify a unity from the multiple. Scission accomplishes this identification by slowing movement long enough to identify a unity.

When applied to the position Georges Peyrol advocates in "The Fascism of the Potato," scission requires each of the groups enumerated in the May '68 "horizontal storm" to see their interests against the larger background of a bourgeois world. It is this work of seeing oneself *placed* and subsequently redefined in the terms of that *place* that Peyrol construes positively; if the "immigrants, women, ecologists, soldiers, prisoners, students, homosexuals, etc." (*FP* 192) each observed themselves against the bourgeois background, the unifying dimension of class would have emerged as key in capitalizing on the promise of May. Such is the faith of Georges Peyrol, a faith that peters out in the transformations of radical leftism toward parliamentary politics in the mid-1980s.

It is in this context that Badiou rethinks the structures for organization and transformation. As I traced in Badiou's *Peut-on penser la politique?* and the reflexive introduction of *Being and Event*, Badiou's search moves from a one cut directly from the multiple to a point of transformation in the form of the void. In this move, expressed in the work Badiou undertakes to establish the void set as the "first existent" and the possibility for any one-result whatever in *Being and Event*, Badiou assigns the work of transformation to a "gap or impasse," a point at which thought finds its limitation and restriction. In this way, the work of one-ing is *indirect* insofar as it consists in a series of operations on the void point, including the application of axioms and installation of relations of belonging and inclusion, and not directly on the inconsistent multiple or mass the void serves to name.

Though, and as A.J. Bartlett and Justin Clemens (2012, 368) have indicated, mathematized ontology serves Badiou's work as a "figure of philosophical fiction," a series of moves necessary to liberate philosophy from the work of thinking being, ontology's positive value is its tools for mapping types of situations—normal, singular, and excrescent—for identifying and prescribing sites of rupture. Just as the elements of constitutive scission can be reduplicated and arranged to describe conditions of suppression or transformation, Badiou's ontological framework in *Being and Event* identifies a maximum of three types of situation: a normal situation, in which everything belongs and is included and thus has no asymmetry; an excrescent situation, in which everything that is included does not belong, the asymmetry of which favors suppression; and a singular situation, in which everything that belongs is not included, an asymmetry which favors transformation. In the ontological Badiou, and on account of the displacement of ontology from philosophy, this mapping is static and formal. An excrescent situation does not generate suppression, but rather only describes the structure of suppression in the language of the multiple; similarly, a singular situation does not burst forth with transformative qualities, but simply describes the structure of transformation according to minimal mathematical and logical relationships of belonging and inclusion. This stasis, prescribed by the conceptual distinction of inconsistent and consistent multiples and the limitation imposed by the axiom of separation, marks the procedures for generating any one from a multiple in Badiou's explicit theory of the multiple, developed in *Being and Event.*

One and Multiplicity in Deleuze and Guattari: Consolidation, Subtraction

In the preceding section I discussed the procedures for one-production apparent in Badiou's oeuvre, including the technical count-procedure appearing in *Being and Event* and the account of scission that marks determination and the identification of existents in his earlier work, and particularly in *Theory of the Subject.* My aim in the present section is to discuss the procedures for connecting one to multiplicity in Deleuze's oeuvre, accounting for the way the continuous multiplicity is shaped in the direction of a conceptual or empirical consolidation.

In the previous chapter, I explained the structure of multiplicity in Deleuze's work, discussing several examples illustrating a consistent commitment to multiplicity defined *between* points, as a system of connection. I discussed the structuring function of multiplicity in the virtual idea and in "1440: The Smooth and the Striated," where a sense of relation between spaces constructed using principles of multiplicity is clarified. The Riemannian principles in play in these examples are, broadly, three. First is the basic distinction of continuous manifolds from discrete ones, where the former are defined by transitions between points. From this assertion, found originally in Aristotle's *Categories*—Riemann develops a second innovation, an account for extending manifolds in multiple dimensions that links points "as a composition of a variability of $n + 1$ dimensions" (Riemann 1873, 15). This transformation allows for the *density* of connection between local points to obtain. The third contribution of Riemann's work is to revise a sense of space such that it need not be evaluated primarily according to a static coordinate space. As Arkady Plotnitsky (2009, 119) explains,

> The key conceptual component of (Riemann's) approach is to define a topologically and geometrically complex space, particularly a curved space such as a sphere, not as a constitution of points, but as a space that could be covered by maps where by it can be treated as locally Euclidean, even though globally the space itself may not be Euclidean.

Per Plotnitsky, Riemann's work contributes to a re-thinking of space to emphasize its "global" behavior according to principles of continuity and multiple-dimensionality, toward the development of so-called "covering maps" that account for links between points in the space. However, this does not exclude the possibility that portions of these maps might be accounted for in coordinate terms. I used the example of my own neighborhood to illustrate this, indicating that though there is an "Euclidean layer" that defines the neighborhood in terms of discrete parcels and measures relationships between these parcels using coordinate information, there are a number of other layers—perhaps more descriptive than the coordinate layer—that are each invented, forged dynamically by an understanding of the neighborhood according to relationships of neighbors to one another, to and from local landmarks, etc. A Riemannian grasp of my neighborhood sees each of these layers and all together as constituting my neighborhood.

It is against this background, that I am now interested to discuss the procedures for one-production in Deleuze's work and in his collaboration with Guattari, though this enterprise is a bit tricky. That is, and unlike the one resulting from the count or the cut in Badiou's work, the one holds no privileged place in Deleuze's and Deleuze and Guattari's work. It is rather understood in conjunction with and as an expression of a local multiplicity. In what follows, I attend to the differential procedures for issuing a pre-individual singularity and a resulting example of a virtual idea—Deleuze's analysis of Marxism as a 'social idea,' in *Difference and Repetition*. I then turn to Deleuze and Guattari's injunction to 'make the multiple' with the figure $n - 1$; I discuss the way this is mobilized by Riemann in the "Hypotheses" and expressed in the example of the war machine in "Rhizome". I supplement this with the imperatives to division and overturned Platonism in *Difference and Repetition*; though they are not properly subtractive, these prefigure this procedure and its later expression. I then address "Example 1" in *What is Philosophy?* to demonstrate both consolidation and subtraction 'in action.'

Consolidation and the Virtual Idea

Attention to the multiplicity structuring the virtual idea, the task of my analysis in Chapter 3, reveals the dynamic connections, structured according to principles of continuity and multiple-dimensionality, constituting that idea. Deleuze's use of the differential element *dx* and its three moments as undetermined, determinable, and determined, mobilizes the multiplicity to consolidate in a virtual idea; it offers, in other words, the conditions by which the internal genesis of an idea is possible.

In the Introduction, I explained that Deleuze's accounts of the virtual idea and the internal genesis it expresses are each offered as a solution to Kant's theory of external conditioning. Kant's solution, offered in the *Schematism*, depends on time "as the formal condition of the manifold of inner sense" (Kant 1998, A138/B177) to facilitate the connection between categories and intuitions necessary to produce an idea. Time is adequate to this task since, as Kant (1998, A138/B177) claims, because it is "a third thing, which must stand in homogeneity with the category on the one and the appearance on the other, and makes possible the application of the former to the latter." Although Kant's argument is exceedingly tidy—time functions as the third term in a syllogism connecting categories and intuitions—it is insufficient because it is *imposed from without* to complete the linkage. As Daniela Voss (2011, 67) explains, Deleuze's disposition toward internal genesis follows the framework established by Solomon Maimon. Rather than depend on a third term to bring categories and intuitions together, Maimon argues that the conditions for the idea are found in the way categories and intuitions are structured. For Deleuze, the principles of internal genesis mobilize the connections evident in a local multiplicity toward a pre-individual singularity. This depends on a (peculiar) interpretation of differential calculus.

However, before arriving at this peculiar interpretation, consider a conventional understanding of differential calculus; the practice determines, over "continuous mathematical magnitude the smallest possible intervals or differentials" (Boyer 1949, 12). These small increments make it possible to identify a value on the continuous magnitude at a particular instant. For example, on a curvilinear function $f(x) = x^2$, it is possible to calculate the average rate of change, or a line secant to the curve between two identified points, and to derive from the average rate the instantaneous rate of change at a given point on the curve, marked by a line tangent to the curve as the rate of change approaches zero.

For a value $x = 2$, the function yields a *y*-intercept: $f(2) = 4$ and an ordered pair designates a point on the curve at (2, 4). To identify the average rate of change, a second pair of coordinates are required. Taking another point on the x-axis an arbitrary distance from 2 described as $2 + h$ (where h marks the change from 2 to the point h), the value of the function at $2 + h$ is evaluated as $f(2 + h) = (2 + h)^2$. Using point-slope formula, generically $m = \frac{y_2 - y_1}{x_2 - x_1}$, the coordinates (2, 4) and $([2 + h], [2 + h]^2)$ determine the slope of the secant line. Expanding $(2 + h)^2$ to $4 + 4\,h + h^2$ and converting it to point-slope formula yields: $m = \frac{(4+4h+h^2)-4}{(2+h)-2}$, which simplifies to $m = \frac{(4h+h^2)}{h}$. The equation is simplified further to $m = 4 + h$ after factoring h from each of the terms. The slope of the line connecting (2, 4) and $([2 + h], [2 + h]^2)$ is thus $4 + h$ and describes the average rate of change over the function.

To determine the instant rate of change at the point (2, 4), the *limit* must be identified. As the value of h, that change from 4 to $4 + h$ approaches a value of zero, the slope of the line tangent to the curve at (2, 4) is calculated taking the limit of the secant line as h approaches zero. The formula $f'(x) = \lim_{h \to 0} 4 + h$ describes the derivative of $f(x)$; as h becomes smaller (or vanishes), the value 4 remains. The slope of the line tangent to the curve at (2, 4) is 4, and describes the instant rate of change at that point.[5]

In this conventional assessment, identifying slopes of secant and tangent lines is operational and determines average and instantaneous rates of change over a function. Deleuze's peculiar interpretation of the differential depends, as he explains, on a "so-called barbaric or prescientific (interpretation) of the differential calculus" (*DR* 170), an interpretation that trades the actual existence of infinitesimals—and their operational function—for their ideal existence.[6] Very basically, this is the claim that *dx* can be put to work without assigning to it any particular mathematical or numeric value.[7]

In his presentation of differentiation in chapter 4 of *Difference and Repetition*, Deleuze explains that the "principle of a general differential philosophy" depends on an interpretation of the differential element *dx* to "(appear) as simultaneously undetermined, determinable, and determination" (*DR* 171). As undetermined, *dx* does not carry a particular numeric value, but rather is the "symbol of difference" in Deleuze's system. It is determinable inasmuch as *dx* may be brought into "reciprocal determination" (*DR* 171) with other differential elements; it is expressed *dy/dx* and does not yet bear a value. Finally, the "principle of complete determination corresponds to the effectively determined," or the point at which a singularity emerges (*DR* 171). This singularity effectively completes the virtual idea initiated in the account of multiplicity and readies it for its expression in space and time.

Deleuze explains that the symbols *dx* and *dy*, which he proffers in order to counter the ~*A* of contradiction, "are completely undifferenciated in the particular and general, but completely differentiated in and by the universal" (*DR* 172). As "undifferenciated," Deleuze insists that these symbols are not attached to empirical or conceptual instances; to be differenciated is to express a multiplicity in a here-and-now instance, to be actualized in space and time. By distinguishing differentiation from contradiction, Deleuze expands the virtual idea according to quantity (a multitude of elements may be in circulation, rather than simply two terms) and quality (a corresponding multitude of relationships may obtain, rather than the narrow relations prescribed in contradiction). Operationally, because differential elements cannot be actualized, they lack "any assignable function" (*DR* 183). In Deleuze's scheme, *dy* and *dx* are abstracted even further from a recognizable mathematical or continuous point; the differential elements of the multiplicity do not carry a particular or specific value. However, these elements are necessary for the definition of an idea-multiplicity.

When Deleuze claims that the differential elements are "completely differentiated in and by the universal" (*DR* 172), this is a gesture to the differential elements in their relation, *dy/dx*. The assertion of "the universal" directs Deleuze's reader to Jean Bordas-Demoulin's claims regarding the differential relation *dy/dx*, and to one of the key contributions of the "pre-scientific" version of the calculus on which Deleuze relies. According to Simon Duffy (2013, 77), Bordas-Demoulin's work sets asides questions of whether the differentials exist, and "instead considers *dy/dx* or 0/0 to be the symptom of a qualitative difference or change of function, insofar as it excludes that which individualizes the function in favor of the universal." Bordas-Demoulin's expression of the universal as 0/0 is peculiar, particularly since any conventional differential relation divided by zero yields an undefined result. What might it mean, then, to "set aside questions of whether the differentials exist"? Henry Somers-Hall offers some orienting comments regarding the commitments by which Deleuze selects Bordas-Demoulin's 0/0. Somers-Hall (2010, 567) explains that in connection to the calculus,

> Deleuze's aim will be to use the calculus to foster an understanding of the transcendental free from all resemblance to the empirical, as it is only once this resemblance has been removed that the transcendental can be seen as the grounds of the generation of the empirical, rather than simply as a conditioning factor.[8]

On Somers-Hall's interpretation, these structures of the differential and its relation are to be interpreted as a set of markers to which empirical results might be mapped in the virtual. Put another way, their interpretation is philosophical, rather than mathematical. If the elements of the differential are relieved of the requirement to communicate precise mathematical values, then the 0/0 is a means of accounting for the transformation of an idea in the virtual; it cannot be undefined, since there are no empirical values associated with it.

When Deleuze says that differential elements are "differentiated" by the universal, he means that differential elements are installed in relation and that it is possible this relation could constitute a qualitative change. Deleuze explains that these differential elements begin as undetermined, but they "must in effect be determined, but reciprocally, by reciprocal relations which allow no independence whatsoever to subsist" (*DR* 183). The multiplicity requires at some level the opportunity for differences to circulate freely. However, these elements carry with them the possibilities for determinability (their installation in relation) and determination (the preparation of this relation to be realized in space and time); they bear the capacity to circulate and be put to work in service of the relation, that second structuring element of a multiplicity.

To realize the possibility of determinability, situating these elements in a strict reciprocal relation is required; this strict reciprocity more-or-less locks the differential elements into a relation, and generates a mode of determination, or a pathway between the elements so related. As such, the second structural feature of the idea articulates the dimensionality of that particular idea. This is expressed symbolically as the relation of *dy/dx*, or "the principle of reciprocal determination" of these elements. In this relation, Deleuze explains, "each term exists absolutely only in its relation to the other: it is no longer necessary, or even possible, to indicate an independent variable" (*DR* 172). This reinforces Deleuze's insistence that the organization of a multiplicity is intrinsic, based on the relations and networks developed between the points of that multiplicity. The dual significance of this second structural feature of multiplicity consists in the determination of points in relation, and confirmation of the idea's continuity. Deleuze says, "In all cases the multiplicity is intrinsically defined, without external reference or recourse to a uniform space in which it would be submerged" (*DR* 183). Multiplicity resists the imposition of structure that is not grown from the reciprocally related elements and the status of this reciprocal relationship vis-à-vis other "neighboring regions" (*DR* 183). The differential relation reinforces the dimensionality and interiority of the idea-multiplicity that marks it as continuous.

A third condition, the "principle of complete determination," accounts for "the effectively determined (values of *dy/dx*)" (*DR* 171). The immediate difference between the second and third conditions of the idea can be expressed as follows: the reciprocal relation of *dy/dx* identifies a place where a change in the function could obtain; the "effective determination" of the value of this relation marks a place where a change in the function has occurred. In other words, the third condition of the idea is the site of an idea's transformation in the direction of its spatio-temporal expression. While the preceding two conditions of the idea are described according to "quantitability" and "qualitability," respectively, Deleuze assigns "potentiality" to the third condition of the idea, suggesting its break in the direction of actualization (*DR* 176). This break is indexed by the presence of a "'preindividual'" singularity (*DR* 176). Simon Duffy (2013, 16) defines the singularity in the context of Leibnizian calculus as "points of articulation where the shape of the curve changes or alters its behavior." As such, the presence of a singularity is crucial for defining the curve; in the conventional example above, a pre-individual singularity would be associated with the infinitesimal point at which the graph of the function breaks from the *x*-axis; this break toward curvature makes it possible to evaluate the rates of change on the function with identifiable (i.e., non-infinitesimal) points. The singularity functions analogously for Deleuze, insofar as it identifies the place at which the idea is prepared to be transformed, to become actual.

Deleuze offers three examples of this virtual Idea, each one affirming the presence of infinitesimal elements, their relationships, and their "potentiality" or their bearing the requisite elements of concrete expression. In the previous chapter, I discussed the relation of the social Idea to Deleuze's framework defining a multiplicity. In this context, I discuss the social Idea's interpretation with respect to the differential. The social Idea develops the elements, relations, and potential in a localized framework, underwritten by the "element of quantitability," or the differential element, the relations determining "qualitability," and the "potentiality of societies" (*DR* 186). Each of these elements, as Deleuze explains, are "established not between concrete individuals but between atomic bearers of labor-power or representatives of property" (*DR* 186). Though the claims concerning localization are perhaps straightforward to grasp (the social Idea is the idea of *this* particular social arrangement and its entailments), this dive to the "atomic" level is not quite as straightforward; large-scale social movements like Marxism are *not* defined as clashes between the bourgeois and proletariat, but rather express traces or micro-tendencies present in any social arrangement. These traces link up, and in doing so lay the ground and limits in which concrete relations and engagements will unfold. Daniela Voss (2013, 198) explains, "Instead of being an abstract transcendent principle, the Idea is not larger than what it determines, that is it is itself determined by the contingent empirical circumstances." Put yet another way, the social Idea poses a question along the lines of *what is meant by "the economic" in this place and at this time*? The answer to this question—the solution—emerges in the concrete expression of economic relationships; however, the solution itself covers over the conditions of its emergence. This requires a bit of imagination to see at work.

Take, for example, my own neighborhood; though I've described this neighborhood previously in terms of covering maps that mark pathways to and from various landmarks, these covering maps describe layers of the neighborhood. To approach the neighborhood from the angle of the social Idea, the answer to the question *What is meant by "the economic" in my Denver neighborhood in 2017?* is *gentrifying*. I can point to a series of concrete indices supporting this answer: rising home prices, long-time residents displaced to more affordable suburbs north and east of the city limits, and economic growth in certain sectors, all offer palpable evidence for this solution, and are each tangled in complex and determining relations with one another. However, the social Idea requires a recognition that even these concrete indices express abstract possibilities somehow 'embedded' in Denver's social arrangement in the early twenty-first century. Put directly, gentrification expresses these embedded traces of opportunity, affordability, displacement, and growth, each mobilized in ways peculiar to this place and time; these traces link up with one another (*growth/displacement, opportunity/affordability, displacement/opportunity, affordability/growth, displacement/affordability, growth/opportunity*) and their nexus identifies their potentiality or the direction of their expression.

The differential and its three moments combine with the framework offered by Riemannian multiplicity to form the structure of the virtual idea. Deleuze describes this structure, saying,

> Just as structure is independent of any principle of identity, so genesis is independent of a rule of resemblance. However, an Idea with all its adventures emerges in so far as it already satisfies certain structural and genetic conditions, and not others. (*DR* 184)

Deleuze reasserts here the nature of the virtual idea according to the critique of the dogmatic image of thought, the project of *Difference and Repetition*, but also in terms consonant with the selection of Riemannian multiplicity and its consequences. The virtual idea resists evaluation by a concept of identity on account of its immanent or intrinsic measure; it "forms a system" by linking certain elements and building pathways between these without appealing to a static unity or its requirements. Similarly, the genetic procedure Deleuze attributes to the differential relation is not governed by pre-existing rules of determination; the results of the differential procedure need not conform to externally imposed norms.

In his introduction to the differential in chapter 4 of *Difference and Repetition*, Deleuze claims, "a great deal of heart and a great deal of truly philosophical naivety is needed in order to take the symbol *dx* seriously" (*DR* 170). To interpret *dx*, Deleuze depends on the tools of a minor tradition of differential calculus for the continued construction of his virtual idea. The resources this minor tradition offers, both in terms of principles of a transcendental (rather than empirical) analysis of *dx* and its success in offering an account of internal genesis are evident. However, we may wonder whether this dependence on such a minor tradition is not also a liability for Deleuze. Consider the claims above that one reason Riemannian differential geometry works in the context of Deleuze's project is on account of its capaciousness. I offered a brief comparison of Riemannian geometry with other geometric approaches which demonstrated that Euclidean and other non-Euclidean geometry can actually be treated as special or highly specific cases of Riemannian space.[9] Deleuze's preferred interpretation of the differential, by contrast, seems to balance on a knife edge. Put another way, one needs far less "philosophical naivety" to accept the principles of continuous multiplicity and immanent measure than one needs to follow a transcendental interpretation of the differential relation. While at present I am not concerned to evaluate the adequacy of the mathematical systems proposed by Badiou and Deleuze for the tasks to which they are assigned, this sharply inverted relation of multiplicity and the differential in Deleuze's virtual idea are deserving of further investigation.

In this section, I have discussed Deleuze's deployment of the differential *dx*, its relation, and transformation toward a pre-individual singularity. This prescribes the work of consolidation in the virtual toward a concrete empirical or conceptual expression. This notion is not particularly straightforward, and though I've worked through an example of the social Idea, additional discussion is required. I return to the work of consolidation below in my analysis of "Example 1" in *What is Philosophy?*, as this draws together both the productive or genetic work of consolidation and the regulative work of subtraction, both key and appearing across Deleuze's work. Before arriving at this example, however, I address Deleuze's and Deleuze and Guattari's use of subtraction, marked with the figure $n - 1$ in regulating the function of any consolidated instance.

Subtraction and 'Making the Multiple'

In Chapter 2, I discussed Deleuze and Guattari's injunction to 'make the multiple' using a tool of subtraction, marked in their work with the figure $n - 1$. I explained this figure as a counter to Lacan's claim that any unity is constructed by addition, and this addition accomplishes the erasure of difference. The subtractive procedure Deleuze and Guattari deploy is designed to *affirm* difference, to see the variety grounding any particular expression. Here, I explore the procedure of subtraction according to a more remote referent found in Riemann's "Hypotheses"; there, and in addition to the $n + 1$ mechanism for expanding manifolds he discusses, Riemann explains that any region of manifoldness may resolve to fewer dimensions and, where possible, to the identification of quanta or points of position in that manifold.

The $n + 1$ is a formula for expansion, for increasing the dimensionality of any manifold. Riemann (1873, 15) follows this presentation with a prescription "to resolve a variability whose region is given into a variability of one dimension and a variability of fewer dimensions." *Resolving* describes the task of isolating a position in any manifold. This procedure is marked $n - 1$; it begins by identifying a given manifoldness and analyzing it using the following parameter:

> Take a continuous function of position within the given manifoldness, which is not constant through any part of that manifoldness. Every system of points where the function has a constant value, forms a continuous manifoldness of fewer dimensions than the given one. (Riemann 1873, 15)

The key notions to take away from this quotation are two: first, that a region of manifoldness is associated with a function that determines the place of points and relations among these points; second, the work of resolving the manifold—in other words, enacting subtraction—requires an identification of points where the function is essentially static, where no change in the function is apparent. This is the case for constant functions, for which the output or result is the same for every input value, or (put another way) where the function registers no change in relation to its input and output values.

It is from these two notions of analysis—the identification of a region of manifoldness and determination of a point of stasis in that manifoldness—that a kind of limitation is enacted. From this point of stasis, Riemann (1873, 15) explains that it is possible to see variability or dimensionality *in relation to this point*, a position that is of fewer dimensions than the original manifoldness. Put another way, though the identification of the point is possible, it is still a point embedded in the series of connections or dimensions that hold for that given region of manifoldness. There may be fewer connections passing through the point than in the fully expanded manifold. Riemann (1873, 15) explains that by this process "the determination of position in the given manifoldness is reduced to a determination of quantity and to a determination of position in a manifoldness of less dimensions." He points out that it may be possible to assign a quantity to the position, but even without that finite identification the work of resolution identifies location and relationships in relation to the resulting dimensionality.

In what follows, I address two examples from Deleuze's work that illustrate these procedures of limitation and subtraction. First, the war machine, discussed in "Rhizome," is an example of subtraction and distribution in a kind of abstract system. The war machine, when understood according to subtraction, affirms Deleuze and Guattari's preference for "a-centred systems," self-organizing and self-regulating systems that resist coordination from above or from without; the static point from which the war machine is revealed is the General. The account of Deleuze's overturned Platonism, a component of *Difference and Repetition*'s argument elevating difference over "the same," offers an early blueprint for subtractive procedure in Deleuze's work, taking the concept *sophist* as its point of constancy or stasis.

A-Centered Systems, War Machine(s)

Subtraction creates sites of connection aligned with rhizomatic thinking, such that "any point on a rhizome can be connected to any other point and must be. This is quite different from trees or roots which fix a point, an order" (*R* 52–53). Subtraction allows for the identification of conditions giving rise to any conceptual or individuated consolidation of a multiplicity. A rhizome is an example of an "a-centred system," which Deleuze and Guattari describe as,

> finite networks of automata in which communication occurs between any two neighbors, where the links or channels are not preexistent, where individuals are all interchangeable, defined only by their state at any given

> moment in such a way that local operations are co-ordinated and the final overall result synchronized independently of any central instance. (*R* 61)

The a-centered system offers a framework for understanding Deleuze and Guattari's preference for a dynamic, self-determining system liberated from centralized control. When the subtractive figure $n - 1$ appears in "Rhizome," it designates a system resisting the imposition of a centralized authority. The war machine, which Deleuze and Guattari identify as a self-organizing, a-centered system, is an example of "a machine multiplicity" that "rejects any centralizing, unifying automaton as 'an asocial intrusion' (15)" (*R* 62).[10] They describe this machine multiplicity as a problem or perplexity:

> The problem of the war machine or firing squad: is a General necessary in order for *n* individuals to fire at the same time; the solution without a General can be found for a acentred multiplicity with a finite number of states and signals of corresponding speed, from the point of view of a war rhizome or a guerilla logic. (*R* 61)

The war machine illustrates the way a multiplicity ought to behave, functioning according to terms it sets itself and resisting terms imposed on it from without. This is accomplished by its ability to fire according to its states and signals, without requiring the command from a general. Deleuze and Guattari explain that the *n* individuals constituting the war machine are "truly always $n - 1$" (*R* 61). In maintaining its self-organization, the war machine also maintains the subtractive position, submitting itself to the logic of the multiple and not the logic of a centralized authority. The work of subtraction and location identified in Riemann's work above, when laid over the example of the war machine, suggests that it is possible to account for the occasion and conditions for a single member of the firing squad's action. However, this is necessarily embedded in and alongside the other members of the grouping and the dynamics (read: occasion and conditions) for their actions.

The a-centered system is consistent with the overriding paradigm of multiplicity Deleuze and Guattari promote. However, the discussion of the a-centered system also reveals subtraction's normative force. That is, a system is not properly rhizomatic, nor does it express a multiplicity when its principle of organization is ceded to a unifying figure or concept; the war machine is no longer a rhizome when it takes its orders from a general. If subtraction is maintained, though, the war machine operates according to its self-coordinating status. The $n - 1$ figure acts as the procedure for distribution and fracture, and ensures multiplicity is not absorbed by a single, unifying instance.

Division and Overturned Platonism

In *Difference and Repetition*, Deleuze describes the procedure for analyzing any empirical or conceptual instance as interrogation. Interrogation finds its model in an interpretation of the activity of division in Plato's *Sophist*, the means and method by which the Eleatic Stranger hunts the sophist. For Deleuze, division offers the means by which it is possible to "(deny) the primacy of original over copy, of model over image" (*DR* 66). In this context, division serves the elaboration of difference "in itself," which is the task of *Difference and Repetition*'s first chapter; for the purposes of this study, it offers a procedure that prefigures the technical work of subtraction prescribed by Deleuze and Guattari in "Rhizome."

In the *Sophist*, Socrates and his interlocutors Theodorus, Theaetetus, and the Eleatic Stranger attempt to identify the true nature of the philosopher. The visitor replies that while *sophist*, *statesman*, and *philosopher* are each distinct, the work of untangling them is a challenge. He proposes to Theaetetus the proper method of hunting the sophist, "by searching for him and giving a clear account of what he is" (Plato 1997, 218c). They begin with the *angler* as their model, attempting to understand the angler by dividing its tasks and activities according to a series of posed differences. For example, they divide angler to the point of identifying him as possessing an art (*technē*), rather than a knack; the angler is a hunter that strikes by hooking, not spearing (Plato 1997, 221b-c). When they arrive at the sophist, they first determine whether the sophist is a layman or an expert (Plato 1997, 221c), and they further determine that the sophist is an expert at *hunting*. The search for the sophist—the hunt for the hunter—affirms some possibilities and rejects others: the sophist takes as their *technē* the "opininative art," which is either committed to truth or to opinion (Plato 1997, 232c). In all cases, however, the rejected possibilities remain in plain view in order to show the trail Theaetetus and the stranger have each taken.

The sophist is determined, not by instances measured against a concept, but (in at least one of the seven divisions) rather by fragmenting the concept 'angler' and holding each of those fragments apart to properly determine the concept. Each division can trace its lineage back to the first division, and the links in the long chain hold fast; the procedure does not collapse or disappear once the desired result is accomplished. Furthermore, these fragments could be held in a strong analogical relation to assess "sophist" and (by degrees) "philosopher." Division (whether applied to *angler* or *sophist*) develops an array of distributed components. The Eleatic Stranger and Theaetetus together weave an elaborate net to nab the sophist.

Daniel W. Smith has assessed the function of dialectic by division, and especially division in the *Sophist* as a model for a Deleuzian search for difference "in itself." Smith (2012, 10) explains that while the other aporetic dialogues, namely the *Phaedrus* and *Statesman*, "move upward toward the 'true lover' or the 'true statesman,'" the movement in the *Sophist* is rather a dive toward the depths. Seeking the sophist uncovers an array of alternatives: sophist, philosopher, angler, forger, and wholesaler, among other distinctions, appear in the work of dividing. This move is also costly, since as Smith (2012, 10) points out, "the method of division can make no appeal to a foundational myth or model, for it is no longer a matter of discerning the true sophist from the false claimant, since *the true sophist is himself the false claimant.*" The work of division in the *Sophist* rejects the ascent to the true altogether, since in each of the seven attempts to discover the sophist it comes eerily close to the philosopher in several respects; the true and the false, as Smith has pointed out, are compromised categories.

The platonism of the *Sophist* thus resists a singular analysis of Platonic dialogues according to a theory of Forms; that is, one does not read the *Sophist* to find a form of the titular character. Further, the work of division enacted by the Eleatic Stranger does not follow the neat line of "What is …?" questioning offered by Socrates in the broader Platonic corpus. Instead of using a controlling concept to evaluate the relationships uncovered in division, the dive to the depths initiated in this process *fractures* a particular idea according to immediate inference and association. In this gesture of fracture, division offers a model for subtraction of the sort I've identified as apparent in Deleuze and Guattari's collaborative analyses. Deleuze's analysis of division in the *Sophist* offers further confirmation of this affinity.

In addition to resisting the internal tendency of dialectic by the "What is…?" question characteristic of the Platonic approach, dialectic by division also resists an external tendency imposed according to the

Aristotelian method of dividing according to genus and species, which requires a middle term to mediate between species and genus (e.g., human as a *rational* animal, where *rational* determines the type of animal the human is). Deleuze claims that the Aristotelian method cannot cope, for example, with the Eleatic Stranger's division of "art into arts of production and arts of acquisition," and raises the subsequent question, "but then why is fishing among the arts of acquisition?" (*DR* 59). There is no middle term to determine fishing as a species of the acquisitive arts; furthermore, to strive for the generality pervading definition by genus and species misses the purpose of division.

Deleuze explains, "Platonic division in no way proposes to determine the species of a genus—or if, rather, it proposes to do so, but superficially and even ironically, the better to hide under this mask its true secret" (*DR* 59).[11] This true secret is the work, as Deleuze repeats, of authenticating and selecting, rather than identifying; the difference is strict, insofar as identification draws an instance under a predetermined concept, while authenticating and selecting are means of testing, "of selecting a pure line from material which is not" (*DR* 60). In other words, selection can only operate when a series of alternatives—some viable, some not—are available to select from.

Deleuze essentially uses a concept or empirical instance as an index that solicits a pursuit of its origins. It is here that division's affinity with subtraction becomes clear. Platonic division's point of departure, he says,

> can therefore be either a genus or a species, but this genus or this large species is understood as an undifferenciated logical matter, an indifferent material, a mixture, an indefinite representing multiplicity which must be eliminated in order to bring to light the Idea which constitutes a pure line of descent. (*DR* 60)

Division can begin its work from either the genus (i.e., animal) or the species (i.e., human); however, division demands a different avenue of analysis, proceeding by the same manner of unfolding modeled in the *Sophist*. The aim is to identify the array of terms in their immediacy and *then* to see the lineage giving rise to the genus, species, angler, or sophist. Deleuze insists that by the method of division, "there are no longer contraries within a single genus, but pure and impure, good and bad, authentic and inauthentic, in a mixture which gives rise to a large species" (*DR* 60). As I discussed above, an imperative accompanies division that requires all the options—the entire array of possible claimants—to be left on the table to select from or to authenticate. Against the Platonic model, Deleuze mobilizes division as follows: "it is a question of making the difference, thus of operating in the depths of the immediate, a dialectic of the immediate" (*DR* 60). Deleuze puts division to work in service of determining difference, which is necessary to understanding multiplicity in his system. In other words, the activity of division and the fragmentation it develops serves to illustrate the superior "mesh" woven in a multiplicity, and to emphasize the complexity of his preferred model for understanding multiplicity (*DR* 182).

The examples of division (in Deleuze's early work) and subtraction (apparent in Deleuze's collaboration with Guattari) each insist on a kind of *regulation*. That is, to accomplish "the task of modern philosophy ...to overturn Platonism" (*DR* 59), the concept must be interrogated and fractured to discover the series and variety of ideas underwriting its expression. Similarly, subtraction functions as a way of *resisting* centralized control and affirming, in the case of the war machine, the possibilities of production interior to any localized multiplicity. Both procedures operate from an identification of a more-or-less static point: division and the work of overturned Platonism require the stable concept of the sophist to proceed, while the war machine describes a self-regulating system subtracted from the figure of the General. My attention now turns to Deleuze and Guattari's work in establishing the diagram of the cogito, appearing in "Example 1" of *What is Philosophy?*, as a site at which both consolidation and subtraction operate.

Consolidation and Subtraction in "Example 1"

In the preceding section, I discussed examples of subtraction at work in examples of the war machine and overturned Platonism. I now turn to Deleuze and Guattari's analysis of the Cartesian cogito in "Example 1" of *What is Philosophy?* Their work with this concept illustrates both consolidation and subtraction together, and functions as an expression of both structure and prescriptive procedure. I am especially keen to show that the deployment of Riemannian notions persists into Deleuze's late collaboration with Guattari, and that these notions continue to function as principles of organization.

Readers of Descartes' *Meditations* are familiar with the procedure by which the meditator arrives at the conclusion, "I think, therefore I am."

Through a practice of hyperbolic doubt, the meditator has effectively cleared away any and all claims about which a doubt might be raised; the purpose of this exercise is the discovery of certainty (Descartes 1984, AT 7.21-22). The first anchor at which the meditator arrives is "I am, I exist" (Descartes 1984, AT 7.25). In an investigation of just what kind of "I" the meditator is, they proceed to realize, first, that they doubt; second, that they think (since doubting is a form of thinking); and third, that thinking is evidence of existence (Descartes 1984, AT 7.26-27). The purpose of this procedure, at least in the immediate context of the Meditations, is to identify the foundation on which subsequent claims of certainty are built, and to eventually accommodate less-than-certain claims in a way that returns the meditator to the world they categorically doubted at the beginning of the project.

Deleuze and Guattari are interested in the cogito in because of their reader's (likely) familiarity with the concept, and because it facilitates an exposition of their conceptual analysis. Deleuze and Guattari argue that this concept bears the conditions of its emergence, since it links doubt, thought, and existence under a single sign. They present the concept in the following diagram in Fig. 4.1, appearing on page 25 of *What is Philosophy?*

The I at the apex stands for the condensed concept, the cogito, while the arcs D, T, and B each mark successive stages of Descartes' process in Meditation Two: doubting, thinking, and being. The subsidiary Is mark points at which some self-reflective awareness occurs and causes the process to change direction (as in, for example, the turn from doubting into thinking, marked by I′).

Deleuze and Guattari to describe their arrangement "in zones of neighborhood or indiscernibility that produce passages from one or to

Fig. 4.1 The cogito

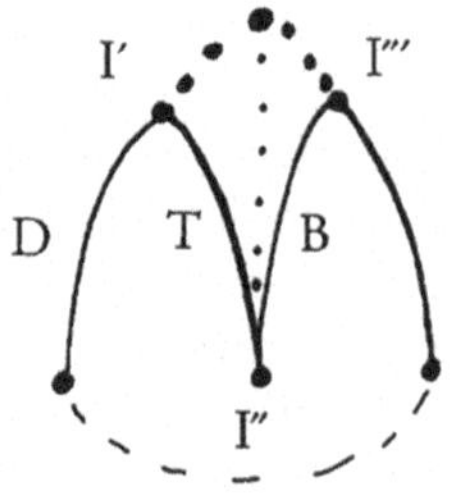

the other and constitute their inseparability" (*WP* 25). Deleuze and Guattari say, "the first zone is between doubting and thinking (myself who doubts, I cannot doubt that I think)" (*WP* 25). The transformation in this first zone is indexed I′. The second zone, "is between thinking and being (in order to think it is necessary to be)" (*WP* 25). This describes the second "hill" of the diagram and connects each of the three components; these components are consolidated or condensed by the apex I.

The diagram supporting Deleuze and Guattari's analysis of the cogito also offers clues for how a reader "makes the multiple" by "subtracting" or bracketing "the unique from the multiplicity to be constituted" (*R* 52). The unique—that is, the mark of consolidation—in the diagram is the point I at the diagram's apex. The work of subtraction requires tracing the underlying links between points in the diagram. Deleuze and Guattari conduct a first subtraction when they claim the apex point "passes through all the components and in which I′ (doubting), I″ (thinking), and I‴ (being) coincide" (*WP* 25). This first subtraction isolates a layer of the diagram expressed by the dotted lines linking these three composite elements and the respective connection of these to the apex I. Deleuze and Guattari describe these components as "intensive ordinates," or positions in the concept at which a qualitative transformation occurs.[12] Doubting, thinking, and being are transformations of kinds of activities, and the ordinates mark where this transformation takes place.

A second subtraction traces the path between these ordinates, allowing Deleuze and Guattari to describe their arrangement "in zones of neighborhood or indiscernibility that produce passages from one or to the other and constitute their inseparability" (*WP* 25). Deleuze and Guattari say, "the first zone is between doubting and thinking (myself who doubts, I cannot doubt that I think)" (*WP* 25). This first zone takes its ordinate as I′. The second zone, they explain, "is between thinking and being (in order to think it is necessary to be)" (*WP* 25). This describes the second "hill" of the diagram, connecting each of the three stages D, T, and B and their respective components I′, I″, and I‴. This description relies on reading for passages or transitions, linkages between points, and an understanding that dimensionality increases by adding available paths to travel between points. Further, the mention of "zones of neighborhood" reinforces the concept's construction as a continuous multiplicity, since neighborhoods describe adjacent subspaces and the passages among these.

This analysis reaches its maximum in Deleuze and Guattari's exposition of the subspaces doubting, thinking, and being. This exposition indicates the divisibility of these explicit spaces into specific, implicit phases. They say,

> … doubt includes moments that are not the species of a genus but the phases of a variation: perceptual, scientific, obsessional doubt … the same goes for modes of thought – feeling, imagining, having ideas – and also for types of being, thing, or substance – infinite being, finite thinking being, extended being. (*WP* 25)

This identification of phases allows Deleuze and Guattari to suggest the way ordinates attract processes. On the basis of this "attraction," it is possible to identify a fourth, speculative subtraction based on the relations of these phases to the ordinate they are assigned. One might imagine a version of the diagram in which the ordinate I′ is itself an expression of relations between *perceptual doubt*, *scientific doubt*, and *obsessional doubt*, where ordinate I″ expresses relations of *feeling*, *imagining*, *ideating*, and where ordinate I‴ accounts for the relations holding among *being*, *thing*, or *substance*. Each of these so-called "implicit phases of a variation" may be traced to smaller or partial instances particular to the locality in which these occur (e.g., each of the times I have—in part or in full—doubted the veracity of my senses constitutes *for me* a version of perceptual doubt).

It is not clear that Deleuze and Guattari are proposing a mechanism for reading the second Meditation as much as they are capitalizing on its most visible contribution to philosophical thought; the cogito is valuable to Deleuze and Guattari insofar as it can be "mechanized" using principles of Riemannian multiplicity and successfully diagrammed. Questions regarding the relation of Deleuze and Guattari's analysis to the cogito in its broader context in the Meditations are not settled, nor do Deleuze and Guattari assess the success or failure of the cogito in the long view (*WP* 27).

Reading Deleuze and Guattari's work with "Example 1" and its attendant diagram makes way for an understanding of the complex relations between a multiplicity and its singular expression; it also allows for an appreciation of the procedure of subtraction. This imperative indicates the necessity of understanding any consolidated instance as expressing a multiplicity, and subtraction offers the procedure for accomplishing this task. While the deep regions of the virtual idea are not explicit in the diagram, I have attempted—in the speculative fourth subtraction proposed above—to characterize what these might consist in.

Summary

The analysis of the diagram in "Example 1," even down to the speculative presence of "phase-variations," expresses a structure consistent with division in Deleuze's early work, with subtraction in the earlier instances of his collaboration with Guattari, and this late work. In Deleuze and Guattari, subtraction takes any unique conceptual or empirical consolidation as a point to examine the continuous multiplicity to which that consolidation is linked. In the preceding discussion of subtraction in the war machine (broadly) and the cogito (specifically), I emphasized that subtraction is essentially *local*, focusing on the concept or empirical consolidation to be interrogated. These figures and their schemes of organization also reinforce structural principles of continuous multiplicity: immanent measure, expansion according to *n*-dimensions, definition and relation according to connected elements. Subtraction is also consonant with earlier figures in Deleuze's work, namely that of division and its service to overturned Platonism. A traditional interpretation of Platonism maintains it as a philosophy minimizing difference in service of conformity to a unifying concept or form. However, Deleuze's suggests that there is a procedure interior to Platonism by which it "has not given up hope of finding a pure concept of difference in itself" (*DR* 59). It is in division and its functioning in Plato's *Sophist* that both elaborate a procedure for determining difference and expressing a subtractive disposition in *Difference and Repetition*.

Conclusion

The work of this chapter has been to discuss procedures apparent in work by Badiou and Deleuze for determining any one from a multiple or multiplicity. As the second of two explicative chapters, I built on the analysis of structure developed in Chapter 3 to emphasize a variety of expressions for "one-ing" a multiple. In some cases, these are prescribed with technical specificity; Badiou's set-theoretic ontology, consisting in the application of axioms to the void set marked Ø, *is* the means by which counting-as-one is possible. I brought Badiou's engagement with Bourbaki to bear, as well as the functioning of axioms of Zermelo–Fraenkel set theory to establish the void as first existent and develop from this first existent sets constructed using relations of *belonging* and *inclusion*.

I explained the turn Badiou makes toward mathematized ontology after having abandoned his purely political—and largely epistemological—efforts of formalization, attending to the account in *Peut-on penser la politique?* And Part IV of the introduction to *Being and Event* in which Badiou registers his exhaustion with traditional forms of materialism. In the ontological turn, readers find a persisting commitment to discover the condition of transformation, though its form is as an "unrepresentable point" that subverts organization applied to it. I compared this ontological approach with a more direct political analysis, appearing in the account of scission Badiou develops in the January 1975 lecture that opens *Theory of the Subject*. There, and working with a particular interpretation of Hegel's dialectic—that of scission, or division, rather than alienation—I suggested that Badiou's approach requires a temporary freezing of dialectical movement, examining any unity according to its component parts.

I then turned to procedures for one-production in Deleuze's work, which I called *consolidation*, beginning with an account of the differential relation. In contrast with a more conventional interpretation of the differential, I described the components of Deleuze's version, which draws on a "pre-scientific" interpretation of the calculus, one largely unconcerned with confirming whether these differentials exist. I described the three key pieces of this program: the element dx, or the quantitability of the differential; the relation dy/dx, or the qualitability of the differential; and the assignation of values to the relation dy/dx, or its potentiality. Together, these three pieces develop the blueprint of a pre-individual singularity, or the ideal conditions required for the concrete or actual inscription of a concept or instance. I discussed the example, which Deleuze offers, of the social Idea to illustrate these elements in action.

Though consolidation is significant in Deleuze's work, it is accompanied by an imperative to *subtraction*, or a procedure by which the consolidated instance is analyzed to discover the multiple underneath. I discussed the technical precedent in Riemann's presentation of the $n - 1$

figure, which Deleuze and Guattari invoke in Rhizome as shorthand for 'making the multiple' (cf. *R* 52). I discussed this procedure in a series of expressions—including interrogation in service of overturned Platonism, its support of an a-centered system like the war machine, and its function in revealing pathways of thought in a diagram of Descartes' cogito—across Deleuze's work and in his collaboration with Guattari. I claimed that subtractive procedure is significant insofar as it marks a consistent commitment in Deleuze's work to maintaining the link between any expression and the multiple from which it is drawn.

The next chapter returns to the series of engagements I outlined in Chapter 2. I insist that a more comprehensive understanding of both Badiou's and Deleuze's respective commitments to multiple, one, and the link between these clarifies the critiques one thinker raises against the other. However, I begin by affirming the presence of subtractive procedure in Badiou's work, thus identifying a second commitment-in-common (in addition to multiplicity) between these opposing programs.

Notes

1. Though Badiou is expressly concerned with the void and its mark Ø here, he has long considered the status of this 'border' between thought and its object. In "Mark and Lack," for example, Badiou is interested "the meta-theoretic function of the zero" ([1969] 2012, 165) and its service in the work of suturing. The early work in "Mark and Lack" situates Badiou in direct conversation with J.A. Miller, whose text "Suture" deploys zero as a means of identifying "the relation of the subject to the chain of discourse" (Miller [1965] 2012, 93). As Paul Livingston (2014) pointed out, in these early inquiries there is no gap in Badiou's work, a position consistent with the texts from the 1970s I am concerned to investigate here.
2. The reference to Weil's comments is available via Jeff Miller's "Earliest Uses of Symbols of Set Theory" website here: http://jeff560.tripod.com/set.html. Miller cites the quotation from Weil's *The Apprenticeship of a Mathematician*, trans. Jennifer Gage (Basel: Birkhauser Verlag, 1992), 114.
3. Bourbaki's presentation of the empty set relies on the universal quantifier ∀ (for all or every object), not on the existential quantifier ∃. In *Theory of Sets* (1968, 72), Bourbaki defines Ø as the value equivalent to a functional relation $(\forall x)$ $(x \notin X)$, which indicates that for every element x, it does not belong to set X; as such, X is empty, since there is no additional

information regarding the contents of set *X*. It is useful to note, perhaps, that Badiou's presentation gets a kind of rhetorical force from the use of the existential quantifier, since the crucial step he is working to establish is the existence of Ø.

4. Badiou's approach to the dialectic and his preference for scission is peculiar. Gabriel Riera (2015), Frank Ruda (2015), and Nina Power (2006), as well as Bosteels (2001) each identify this tendency in Badiou's approach.
5. This example is adapted from the Khan (2015), "Calculating Slope of Tangent Line Using Derivative Definition," available online at: https://www.khanacademy.org/math/differential-calculus/taking-derivatives/derivative_intro/v/calculus-derivatives-2-new-hd-version.
6. In a (brief) description of the transformation differential calculus undergoes in the nineteenth century, Manuel DeLanda (2012, 223) explains this transformation as a consequence of the belief that "*all of mathematics had been reduced to logic* in the late nineteenth century: first the differential calculus was reduced to arithmetic—the concept of infinitesimals was replaced by the notion of limit and the latter reduced to that of number—and later on arithmetic was reduced to set theory." The transformation of mathematics along these lines is noted, not only by DeLanda, but also by Daniel W. Smith ([2003] 2012). It is of key significance for Deleuze's work inasmuch as it facilitates a distinction between the royal or major science reflected by axiomatics, in contrast to the nomad-minor science of problematics that Deleuze depends on across his oeuvre. In *DR*, Deleuze addresses this transformation briefly on pp. 176–177.
7. Deleuze uses Leibniz's notation for the differential relation in his work. Morris Kline (1967, 375–376) explains that Leibniz's notation, "though suggestive of what takes place, can be misleading, for the instantaneous rate of change of *y* with respect to *x* is not a quotient but rather the limit approached by the quotient k/h)." In subsequent discussion, Klein substitutes the Newtonian notation $\dot{y}$ for the *dy/dx* relation.
8. See also Henry Somers-Hall (2010, 566–570) and Simon Duffy (2013, 47–88). Somers-Hall's comments on the calculus offer helpful orientation toward the broad ideas. Duffy's work situates these "prebarbaric" commitments of the calculus in a broader conversation about the Kantian precedent to which Maïmon (chiefly) reacts, and which avails Deleuze of the resources of Maïmon, Bordas-Demoulin (contra. Leibniz in particular), and Wronski.
9. See Uta Merzbach and Carl Boyer (2011), *A History of Mathematics*, 3rd ed. This insight also appears in Penrose (2004, 138).
10. The parenthetical reference (15) is to Rosenstiehl and Petitot (1974), "Automate asocial et systems acentrés," *Communications* 22: 45–62.

11. In a footnote appended to this claim in *DR*, Deleuze refers his reader to Aristotle's critique of division in the *Prior* and *Posterior Analytics*; he returns this critique with a direction to *Statesman* 266 b-d, in which "the determination of species is merely an ironic appearance and not the aim of Platonic division" (*DR* 312n20).
12. Manuel DeLanda (2002, 25) describes an intensive property as "not so much one [that] is indivisible but one which *cannot be divided without involving a change in kind*." He illustrates this with an example of a volume of water, the temperature of which is transformed when that volume is heated from below: "while prior to heating," DeLanda explains, "the system is at equilibrium, once the temperature difference is created the system will be away from equilibrium, that is, we can divide its temperature but in so doing we change the system qualitatively" (2002, 25). This example is designed to demonstrate the transformation of an intensive property—temperature, in this case—results in a change to the *system* (i.e., from equilibrium to disequilibrium), not just to the volume of water being heated.

References

Aristotle. 1984. *Metaphysics*. Translated by W.D. Ross. In *The Collected Works of Aristotle*, vol. 2, edited by Jonathan Barnes, 1552–1728. Princeton, NJ: Princeton University Press.

Badiou, Alain. 1985. *Peut-on penser la politique?* Paris: Seuil.

———. 2005. *Being and Event*. Translated by Oliver Feltham. London: Continuum. Originally *L'être et l'événement*. Paris: Éditions du Seuil, 1988. (Cited in text as *BE*)

———. 2009. *Theory of the Subject*. Translated and with an Introduction by Bruno Bosteels. London: Continuum. Originally *Théorie du sujet*. Paris: Éditions du Seuil, 1982. (Cited in text as *TS*)

———. [1965] 2012. "The (Re)Commencement of Dialectical Materialism." In *The Adventure of French Philosophy*, 133–170. Translated by Bruno Bosteels. London: Verso.

———. [1969] 2012. "Mark and Lack." Translated by Zachary Luke Fraser with Ray Brassier. In *Concept and Form, Volume One: Key Texts from the* Cahiers pour l'Analyse, edited by Peter Hallward and Knox Peden, 151–158. London: Verso.

———. 2012. "The Fascism of the Potato." In *The Adventure of French Philosophy*, 191–201, edited and translated by Bruno Bosteels. London: Verso. Originally Georges Peyrol. "Le fascisme de la pomme de terre." In *La situation actuelle sur le front de la philosophie*, 42–52. Paris: François Maspero, 1977. (Cited in text as *FP*)

Badiou, Alain, and Gilles Haéri. 2016. *In Praise of Mathematics.* Translated by Susan Spitzer. Cambridge, UK: Polity Press. Originally *Éloge des mathématiques*, Paris: Flammarion, 2015.

Badiou, Alain, and Tzuchien Tho. 2007. "*The Concept of Model* Forty Years Later: An Interview with Alain Badiou." In *The Concept of Model: An Introduction to the Materialist Epistemology of Mathematics*, 79–104, edited and translated by Zachary Luke Fraser and Tzuchien Tho. Melbourne: re.press.

Bartlett, A.J., and Justin Clemens. 2012. "II: Neither/Nor." *Critical Inquiry* 38 (2): 365–380. https://doi.org/10.1086/662747. Accessed 1 September 2013.

Bosteels, Bruno. 2001. "Alain Badiou's Theory of the Subject: Part I. The Recommencement of Dialectical Materialism?" *Pli* 12: 200–299.

Bourbaki, Nicolas. 1968. *Elements of Mathematics: Theory of Sets.* Paris: Hermann.

Boyer, Carl. 1949. *The History of Calculus and Its Conceptual Development.* New York: Dover.

Brassier, Ray. 2005 "Badiou's Materialist Epistemology of Mathematics." *Angelaki* 10 (2): 135–150.

DeLanda, Manuel. 2002. *Intensive Science and Virtual Philosophy.* Continuum Impacts Series. London: Continuum.

———. 2012. "Deleuze, Mathematics, and Realist Ontology." In *The Cambridge Companion to Deleuze*, edited by Henry Somers-Hall and Daniel W. Smith, 220–238. Cambridge, UK: Cambridge University Press.

Deleuze, Gilles. 1994. *Difference and Repetition.* Translated by Paul Patton. New York: Columbia University Press. Originally *Différence et repetition.* Paris: Presses Universitaires de France, 1968. (Cited in text as *DR*)

Deleuze, Gilles, and Félix Guattari. 1981. "Rhizome—Introduction." Translated by Paul Patton. *I & C* 8 (Spring): 49–71. Originally *Rhizome: Introduction.* Paris: Les Éditions de Minuit, 1976. (Cited in text as *R*)

———. 1994. *What Is Philosophy?* Translated by Hugh Tomlinson and Graham Burchell. New York: Columbia University Press. Originally *Qu'est-ce que la philosophie?* Paris: Les Éditions de Minuit, 1991. (Cited in text as *WP*)

Descartes, Rene. 1984. *Meditations on First Philosophy.* In *The Complete Works of Descartes*, vol. 2. Translated by John Cottingham, Robert Stoothoff, and Dugald Murdoch, 1–62. Cambridge, UK: Cambridge University Press.

Duffy, Simon. 2013. *Deleuze and the History of Mathematics: In Defense of the "New".* Bloosmbury Studies in Continental Philosophy. London: Bloomsbury.

Feltham, Oliver. 2008. *Alain Badiou: Live Theory.* London: Continuum.

Hegel, G.W.F. 2010. *The Science of Logic.* Translated by George di Giovanni. Cambridge, UK: Cambridge University Press.

Kant, Immanuel. 1998. *Critique of Pure Reason*. Translated by Paul Guyer and Allen W. Wood. Cambridge, UK: Cambridge University Press.

Khan, Sal. 2015. "Calculating Slope of Tangent Line Using Derivative Definition." Khan Academy. https://www.khanacademy.org/math/ap-calculus-ab/ab-derivative-intro/ab-defining-derivative/v/calculus-derivatives-2-new-hd-version. Accessed 27 May 2015.

Kline, Morris. 1967. *Mathematics for the Non-mathematician*. New York: Dover.

Livingston, Paul. 2014. "Formalism and the Real: Ontology, Politics, and the Subject." Lecture, Pittsburgh Summer Symposium in Contemporary Philosophy, Duquesne University, Pittsburgh, August 6.

Lucretius. 1975. *De Rerum Natura*. Translated by W.H.D. Rouse. Loeb Classical Library, edited by Jeffrey Henderson. Cambridge, MA: Harvard University Press.

Merzbach, Uta, and Carl Boyer. 2011. *A History of Mathematics*, 3rd ed. Hoboken, NJ: Wiley.

Miller, Jacques-Alain. [1965] 2012. "Suture (Elements of the Logic of the Signifier)." Translated by Jacqueline Rose. In *Concept and Form, Volume One: Key Texts from the* Cahiers pour l'Analyse, edited by Peter Hallward and Knox Peden, 91–102. London: Verso.

Miller, Jeff. 2017. "Earliest Uses of Symbols of Set Theory and Logic." Last updated 23 June 2017. http://jeff560.tripod.com/set.html. Accessed 12 July 2017.

Penrose, Roger. 2004. "Riemann Surfaces and Complex Mappings." In *The Road to Reality: A Complete Guide to the Laws of the Universe*, 135–152. New York: Vintage.

Plato. 1997. *Sophist*. Translated by John M. Cooper. In *Plato: Complete Works*, edited by John M. Cooper, 235–293. Indianapolis, IN: Hackett Publishing Company.

Plotnitsky, Arkady. 2009. "Bernhard Riemann's Conceptual Mathematics and the Idea of Space." *Configurations* 17 (2): 105–130. https://doi.org/10.1353/con.0.0068. Accessed 12 June 2013.

Power, Nina. 2006. "Towards an Anthropology of Infinitude: Badiou and the Political Subject." *Cosmos and History: The Journal of Natural and Social Philosophy* 2 (1): 186–209.

Riemann, Bernhard. 1873. "On the Hypotheses Which Lie at the Bases of Geometry." Translated by W.K. Clifford. *Nature* VIII (183–184): 14–17, 36–37.

Riera, Gabriel. 2015. "The Question of Art: Badiou and Hegel." In *Badiou and Hegel: Infinity, Dialectics, Subjectivity*, edited by Jim Vernon and Antonio Calcagno, 77–102. Lanham, MD: Lexington Books.

Ruda, Frank. 2015. "Badiou with Hegel Preliminary Remarks on A(ny) Contemporary Reading of Hegel." In *Badiou and Hegel: Infinity, Dialectics,*

Subjectivity, edited by Jim Vernon and Antonio Calcagno, 105–122. Lanham, MD: Lexington Books.

Smith, Daniel W. 2012. "The Concept of the Simulacrum: Deleuze and the Overturning of Platonism." In *Essays on Deleuze*, 3–26. Edinburgh: Edinburgh University Press.

Somers-Hall, Henry. 2010. "Hegel and Deleuze on the Metaphysical Interpretation of the Calculus." *Continental Philosophy Review* 42: 555–572. https://doi.org/10.1007/s11007-008-9120-2. Accessed 12 July 2013.

Voss, Daniela. 2011. "Maïmon and Deleuze: The Viewpoint of Internal Genesis and the Concept of Differentials." *Parrhesia* 11: 62–74. http://ParrhesiaJournal.org/parrhesia11/parrhesia11_voss.pdf. Accessed 14 November 2011.

———. 2013. *Conditions of Thought: Deleuze and Transcendental Ideas.* Edinburgh: Edinburgh University Press.

Zermelo, Ernst. 1977. "Investigations in the Foundations of Set Theory (1908)." Translated by Stefan Bauer-Mengelberg. In *From Frege to Gödel: a Sourcebook in Mathematical Logic*, edited by Jean van Heijenoort, 199–215. Cambridge, MA: Harvard University Press.

CHAPTER 5

Re-engagements

My work in the previous two chapters elaborates the ideas of multiple appearing in Badiou's work, the commitments to and nature of multiplicity across Deleuze's oeuvre, and the procedures governing the relationship of multiple and multiplicity to one in their respective programs. The path I've cut through this landscape has emphasized very basic continuities holding between the two systems: their shared affirmation of multiple as the ground or foundation of being and the use of mathematic structures to organize multiplicity are hallmarks of this "moment" in ontological inquiry. This chapter begins by recognizing one additional continuity, evident in the presence of a subtractive procedure in Badiou's work and one on the order of the $n - 1$ that "makes the multiple" for Deleuze and Guattari.

This procedure, which I label *general subtraction*, functions as a "divestment procedure" that reveals the presence of inconsistency in any situation (Bartlett and Ling 2014, 3). For the sake of clarity, I distinguish this *general subtraction* from the procedure Badiou outlines concerning the identification and assertion of any truth described in the Gamma Diagram; I label this move of subtraction *special subtraction* and draw out the differences between the special and general accounts in Badiou's work. This alignment indicates the overlapping recognition of a fundamental procedural necessity in these often-opposed programs; I insist on the presence of a continuous procedure, in spite of Georges Peyrol's claims in "The Fascism of the Potato" that subtraction fails to serve multiplicity (as Deleuze and Guattari prescribe) and only reinforces

B. Vartabedian, *Multiplicity and Ontology in Deleuze and Badiou*,
https://doi.org/10.1007/978-3-319-76837-3_5

the presence of the one and multiple (*FP* 199). As I discuss below, however, subtraction is only possible in the *ontological* register; there are important ways in which subtraction differs from the technique of scission Peyrol proposes. These differences are another way of marking the "periodization" internal to Badiou's own thought, the move from *political epistemology* to *ontology*.

I address two frameworks in which Badiou's subtractive procedure—when it is applied apart from the rich conceptual resources of the ontology developed in *Being and Event*—create liabilities for his overall program. The first emerges in Badiou's critique of Heidegger in *Briefings on Existence*; when Badiou applies subtraction to make his ontology conversant with a Heideggerian one, he empties his program of the distinctive features that establish it as a genuine alternative to Heidegger's program. Second, and along similar lines, I return to Kant's notion of the manifold and Badiou's critique in "Kant's Subtractive Ontology." In this brief text, and one in which—again—Badiou poses his ontological project as an alternative to the Kantian program, Badiou ends up positing his ontology in a roundabout way as an "ontology of presence", an alternative he works in the early stages of *Being and Event* to strenuously avoid. The solution to these problems is found in the tricky balance of affirming the specific, mathematical and mathematically-derived concepts that populate Badiou's ontology (on the one hand) and affirming the restriction of ontology to mathematics (on the other hand).

I opened this text by claiming a conversational approach occasions a more comprehensive look at the programs under discussion and, in neutralizing rhetoric associated with more combative or polemical critiques, offers a finer-grained understanding of the relation of Deleuze's theory of multiplicity to Badiou's. The work of the preceding two chapters completed the comprehensive look, and the second task of this chapter is to approach once again the relation of the two programs. I begin with the discussion precipitated by "Rhizome" and its various replies, as well as the assessment by Deleuze and Guattari of Badiou's program in *What is Philosophy*? Returning to this set of exchanges with an expanded understanding of structure and procedure reinforces the sense in which Deleuze and Guattari's "Rhizome" applies a form of the multiple anathema to the programs they critique. Further, attention to structure and procedure show Badiou's dependence on an ontologically-static frame of reference in which procedure operates, a stasis that has consequences for accounting for the new or novel in his work.

I then take up Badiou's claim that Deleuze's *oeuvre* is populated with 'monotonous productions,' interpretations of singular or unique instances that serve his overall theoretical trajectory, apparently at the cost of this uniqueness or singularity. With analysis by Simon Duffy (2013) on Deleuze's use of mathematics, I demonstrate that this is only a liability if one is committed to the *consequent* of the radical thesis, that mathematics = ontology. This also serves as corrective to the discussion, apparent in Badiou's "One, Multiple, Multiplicities," that his preferred mathematical paradigm—and only *his* preferred paradigm—adequately mobilizes Riemann's insights.

Subtractive Analysis: $n - 1$ and a 'General Subtraction' in Badiou

In the continuous multiplicities from which Deleuze and Guattari's key concepts are built, $n - 1$ acts as a procedure, distributing the localized multiplicity to avoid its total absorption by a single, unified interpretation or instance. In "Rhizome," Deleuze and Guattari insist that the work of subtraction, required for the unveiling of any multiplicity, cannot simply be *declared*; it must instead be an act of analysis (*R* 52). In the preceding chapter, I discussed examples from Deleuze and Guattari's work that showed a commitment to the presence of one understood only in relation to the multiplicity grounding it; per Deleuze and Guattari, it is only on account of this subtracted status that the one maintains any presence with respect to the multiple (*R* 52).

Because it need only maintain a determined, continuous multiplicity in relation to its unique expression, the procedure by which the multiple is revealed and is maintained can be the same procedure. To illustrate these ideas, I discussed several examples from Deleuze and Guattari's work, emphasizing the "a-centred system" any rhizomatic multiplicity expresses. The diagrammatic analysis of the cogito in "Example 1" shows, for example, that the cogito is not simply posited in the single point, "I think, therefore I am." Rather, it is accompanied by its constitutive elements of *doubt, thought*, and *existence*; these elements form 'zones of indiscernibility' among which pathways are etched, and a map tracing the concept's construction appears. In this context, subtraction reinforces the complexity attending any single expression, and urges the analyst to seek the conditions and connections producing it.

As I show in this section, Badiou's subtractive system is the work of disposing ontological thought to its seat in the void, to the discovery of thought's fundamental limit at the boundary of inconsistent multiplicity. In this way, it serves a similar function to the $n - 1$ procedure in Deleuze and Guattari's work; it does not include a reinforcing or regulative function, however, and is distinguished from Deleuze and Guattari on this account. The presence of a "descriptive" subtraction, one that identifies the multiple-character of any ontological construction, also forms a continuity between Badiou's program and that of Deleuze and Guattari. However, I demonstrate ways in which the subtractive impulse, when it is divorced from the rich conceptual vocabulary of Badiou's ontology, becomes a liability and reveals keen limitations of Badiou's account; this so-called 'meta-ontological' subtraction is a move Badiou deploys when situating his program as an alternative to other ontological proposals, namely those of Heidegger and the "subtractive ontology" Badiou reads in work by Kant.

Special and General: Ontology, Subtraction

Some terminological precision around the notion of subtraction in Badiou's work is required. One finds subtraction to refer to the process by which a truth is analyzed (a process described by the Gamma Diagram); one also finds an insistence on subtraction as a requirement for maintaining the status of being as multiple. In order to fix this terminology a bit more, I propose a distinction of *special subtraction* from *general subtraction*, a distinction Badiou applies to ontology in *Briefings on Existence*. There, Badiou contends that against the background of the Cantorian identification of in/consistent multiplicities, it is possible to "move from 'special ontology,' which still links the multiple to the metaphysical theme of representing objects, numbers and figures, to 'general ontology,' which sets the free, thoughtful apprehending of multiplicity as such as the basis and destination of mathematics" (*CT* 40–41). Special ontology, then, requires multiplicity to re-present everything that is; it imposes a requirement on multiplicity-qua-being to account for the world as it appears in its specificity and complexity. As I showed in the previous chapter, Badiou's ontology is not capable of this. Indeed, it resists this impulse by forming out of the void consistent multiples, or *structured situations*, that describe the arrangement of terms by relations of belonging and inclusion. As such, Badiou's ontology is a candidate for *general ontology*, the work of thinking or organizing pure multiplicity as multiplicity; the requirements of general ontology are minimal, insofar as the so-called "apprehension of multiplicity" is the only task it is assigned. The distinction of special and general is thus founded on a kind of limitation of scope; where *special* designates work devoted to a specific object or process, general designates a disposition toward the foundation of all thought in pure multiplicity. Special ontology, in other words, reads multiplicity toward the world as it appears; general ontology, by contrast, reads multiplicity in the direction of inconsistent multiplicity. This distinction can be applied to subtraction in Badiou's work, an application that clarifies—and perhaps fixes—its use in a more helpful way. *Special subtraction* appears in Badiou's work as the fourfold framework mobilizing the philosophical investigation of truths, while *general subtraction* describes the retroactive procedure of identifying the multiple in any structured situation.

In the opening paragraphs of "On Subtraction," an essay appearing in the collection *Conditions*, Badiou (2008, 113) defines subtraction as "an act ... the paragon of the act, the act of a truth, the one through which I come to know the only thing one may ever know in the element of the rea: the void of being as such." Subtraction is thus a means of access to the only accessible element of being qua being—that of the void. The diagram traces its presence through a post-evental process, that of discerning the emergence of a truth. Subtraction is associated with the "Gamma Diagram," a map of sorts, layering "the logical figures of subtraction onto an ontological distribution" (Badiou 2008, 122). Ontological categories including the *one-more*, the *finite*, the *infinite*, and the *one-less* account for the status of a truth relative to the situation in which it emerges; its trajectory is accounted for in logical terms that map (respectively) any truth's status as undecidable, indiscernible, generic, and unnameable. It is this logical apparatus that exposes what was ontological in philosophical—or *meta-ontological*—terms.

Here subtraction is a method of separation acknowledging the singular status of a truth against familiar or established standards of evaluation: the undecidable indicates any truth resists norms or laws; the indiscernible is any truth's resistance to marking (Badiou 2008, 115); the generic describes the character of any truth in its infinity, or in its capacity to take on "an excess of determinations" which "engenders an effect of indeterminacy" (Badiou 2008, 118); the unnameable is that truth's resistance to familiar or ready expressions. In the context of this special subtraction, these resistances are facilitated by the void's presence in the situation that generated the truth: the event, itself a result of the void's irruption in any consistent multiple. This form of subtraction shares the impetus toward localization apparent in a so-called special ontology. That is, it takes a specific and narrow focus on which to operate (truth) and opens this narrow focus to a broader investigation. Though Badiou would not affirm a special ontology per se, distinguishing a special form of subtraction recognizes the passage of the void through a very narrow framework. It also situates special subtraction as a *liminal* procedure, exposed on an ontological side such that it can be grasped in language appropriate to a mathematized ontology, and on a logical side, with language appropriate to the broader conditions—art, science, love, politics—in which a truth might appear.

The subtraction documented by the Gamma Diagram is not the only way of 'siting' the void. A.J. Bartlett and Alex Ling define a subtraction that moves in a different direction. Bartlett and Ling (2014, 4) say,

> Once an object is divested of (or 'subtracted from') everything that goes into making it a 'unique thing' – once we isolate it from its context and strip away each and every one of its qualitative determinations – what remains is essentially a multiple of multiples. This pure multiple remainder is literally the being of the object, the elementary 'there is' underlying everything that 'is there.'

In my discussion of Badiou's count-procedure in the preceding chapter, I explained that the void is the point at which thought grasps the proper name of being; the void simultaneously maintains the nature of being as inconsistent and founds ontology as the construction of consistent multiples. Bartlett and Ling's characterization of what I call general subtraction begins from things, a passage possible from the analysis in *Logics of Worlds*.[1] Bartlett and Ling emphasize a "divestment procedure" that drives inquiry toward the multiple; insofar as ontology is a theory of pure multiplicity, this peeling away serves as a reminder of the content of any situation or thing as originally multiple. Their presentation is also consistent with an early claim of Badiou's in *Being and Event*, where he insists, "The multiple is retroactively legible therein as anterior to the one, insofar as the count-as-one is always a result" (*BE* 24). Following the decision that the One is not, Badiou's analysis of any consolidated or counted situation reveals the prior and persisting presence of the multiple as the being of that consolidated or counted situation.

General subtraction is a mechanism that maintains the disposition of ontology toward multiples (not ones or multiples counted-as-one) and toward maintaining the character of being qua being as the multiple of multiples. In *Briefings on Existence*, Badiou explains that the subtractive disposition must continually be pursued, at least until thought arrives at the limit posed by inconsistency: "so long as a multiple (a being)," Badiou says, "is not a manifold of multiples, subtraction has to be maintained until the very end" (*CT* 37). It suggests a shift from any consolidated situation to the elements that situation is composed of, and then to a subsequent recognition of the source of this composition in the void set. As such, subtractive analysis is motivated by an imperative to seek inconsistency at the foundation of ontological operation. Though my work in the preceding chapter demonstrated the construction of counts from the void (and so moved *prospectively*), a requirement of general subtraction or divestment maintains a link to the void founding any ontological procedure—including one as complicated as the truth-procedure expressed in the Gamma Diagram. As such, general subtraction is *local* insofar as it describes the pathway to the void and inconsistent multiplicity from any particular, consolidated situation.

This radical localization supports a theory of *ontological rank*, appearing in *Logics of Worlds* and accounting for the relationship of any world—which Badiou describes as "the place in which objects appear" and the contingent relations between elements in that "closed ontological set" (2009, 598)—to multiplicity.[2] Badiou (2009, 112) explains, "Ultimately, it is clear that every thinkable being is drawn from operations first applied to the void alone. A multiple will be all the more complex the longer the operational chain which, on the basis of the void, leads to its determination. The degree of complexity is technically measurable: this is the theory of 'ontological rank'." The theory of ontological rank maintains the disposition of general subtraction I've elaborated here as Badiou's theory of appearances adds complexity to the ontology of situation in *Being and Event*. While their qualitative determinations are accounted for inside the world in which these appear, Badiou's theory of ontological rank keeps an appearance operationally linked to the void and so to inconsistent multiplicity; their quantitative status is determined by the connection each world maintains to the void. Lindsey Hair

(2006, 279n4) explains that the complexity of any world is determined by the originary link to inconsistency, and reinforces its locality:

> ...since there is no whole of Being, there can be no single scale upon which multiple being can be ordered – there are multiples whose construction does not intersect with other multiples outside the single shared foundational set of the void. This cancels the possibility of any global uniformity or categorization of beings: identifications and relations are always local.

The extremity of closure Hair suggests (and a more detailed analysis of *Logics of Worlds* would offer) supports a radically-localized subtractive procedure. To be sure, understanding the transition from *world* to *void*, from *quality* to *quantity*, depends in part on understanding the mathematical links Badiou develops between set-theoretic ontology and category-theoretic ontology. However, and for the present purposes, the significance of this discussion is to indicate the possibilities for reading general subtraction from complex appearances to the void. These examples suggest that any general mode of subtraction is a *disposition* in Badiou's ontology, that is, a turning of inquiry toward inconsistent multiplicity at its foundation and an accompanying unbinding of any situation in search of the multiple of multiples. This unbinding moves from the situation, to the elements collected in the count-procedure, to the source of composition in the void set; the count does not go to work *directly* on inconsistent multiplicity, only that exposure of being in the void.

The assessment of structure in this work's Chapter 3 demonstrated the differences in kind between the multiplicities at work in Badiou's and Deleuze and Guattari's respective work. This difference must be maintained to avoid any elision between the two programs; however, the analysis of general subtraction in Badiou offers a perhaps unexpected continuity between his program and that of Deleuze and Guattari's. General subtraction, like the $n - 1$ procedure, begins from a situation an instance of consolidation, analogous in its location to the conceptual or empirical expression in Deleuze and Guattari's account. The assessment of procedure in the preceding chapter makes it possible to see these in analogous positioning, insofar as the one of situation (in Badiou's work) remains composed of and tethered to inconsistent multiple by way of the void, and the one of concept or instance (in Deleuze and Deleuze and Guattari's work) is tethered to the continuous multiple it expresses.

In Deleuze and Guattari, subtraction takes a unique conceptual or empirical expression and suspends it to discover the continuous multiplicity from which that expression emerges. Subtraction, when exercised on the diagram associated with "Example 1"—a task I took up in Chapter 4—indicates that there is no need to look beyond this identified multiplicity for its source. There are a number of elements a continuous multiplicity mobilizes: intensive ordinates mark the qualitative changes in the process, linkages between neighboring points trace the path between ordinates, and the ordinates themselves serve as "attractors" for related phases or variation. This example reinforces principles associated with Deleuze's use of Riemann's continuous manifolds, including immanent measure, expansion according to *n*-dimensions, definition according to elements and relations between elements. It also suggests that all subtraction for Deleuze and Guattari is *local*, limited to the concept or empirical expression and the multiplicity to which it is tethered.

Badiou's subtraction, though general in its disposition toward inconsistent multiple at the foundation of any situation, is also local insofar as it begins (at least) from an ontological situation, unbinding layers of structure to reveal the void at the foundation of that situation. Subtraction reduces the situation into a more primitive procedure of forming-into-one, and further to its most minimal presentation in the void set. Beyond the border presented in the void set, and inaccessible to thought, is the inconsistent multiple; subtraction *suggests* the presence of inconsistent multiple to thought in the void set, but the inconsistent multiple is beyond the reach of mathematical structure. However, unlike Deleuze and Guattari's subtractive procedure, which both makes and maintains a continuous multiplicity, the subtractive procedure Badiou uses to uncover multiplicity finds its limit at the void set. There is no access, at least with the tools of ontology, to inconsistent multiplicity beyond the void set. However, general subtraction maintains a 'dispositional link' toward the void set and inconsistent multiplicity beyond it.

The procedure of general subtraction I've attributed to Badiou's ontological work *is not* continuous with his earlier, politically-motivated work. In Chapter 4, I took up a discussion of scission from the 1975 lecture that opens Badiou's *Theory of the Subject*. There, he reviews the Maoist principle "One divides into two" using a dialectical matrix emphasizing *division* and not alienation; in other words, Badiou's chosen interpretation of the dialectic is to read a single moment of the dialectic according to its two preceding and constitutive moment, rather than the movement of alienation and sublation that moves the dialectic forward. I explained that scission relies on a kind of *freezing* or *pausing* to discover the "minimum primary differential" distinguishing determinate being and its qualitative opposite, both of which contribute to the determination of any existent. Though it is implied that each of the moments under investigation carry the heritage of precedent dialectical moves to their particular moment of analysis, Badiou's interest in scission is simply to identify the condition of difference *and from that point* initiate the work of combination and recombination toward a structural analysis of transformative force and repressive splace. Put another way, scission, for Badiou, reads for the moment of contradiction and no further.

Badiou's ontology is governed by a similar imperative insofar as the point sought is that of the void. However, and as I explained in Chapter 4, the void point is a *complement* of any ontological structure, more-or-less successfully covered over by the actions of the count, and ultimately discoverable by the divestment of general subtraction. The language of minimum primary difference is abandoned in the ontological work in favor of a deep point of rupture or fracture. The dictum transforms with the ontological turn from "One divides into two" to "One divides into multiple." Though the impulse is the same, Badiou's ontological analysis—because it is loosed from the requirement of representing anything (it is not, in other words, a form of 'special ontology'), can thus accomplish a subtractive inquiry that moves through strata of structure to discover a link to inconsistent multiplicity in the void point. There is also a way in which the procedure described in scission differs from the analysis of division I attributed to Deleuze's account of overturned Platonism in Chapter 4. While Badiou's assessment of division reads until the place of contradiction is uncovered, the work of division and overturned Platonism requires a "deformation" of a concept at all of its natural joints, until the concept can be divided no further. There is no pursuit of contradiction in this process, only differentiation (in a nonmathematical sense), pursued to identify all that underwrites it.

In the context of Badiou's *ontological* project—the project governed by mathematical concepts and apparatus—general subtraction keeps the void and inconsistent multiple in view. This is significant as the movement from being to appearing gains considerable complexity. However, the move Badiou makes in the "The Question of Being Today" and again in

"Kant's Subtractive Ontology" situates subtraction in a meta-ontological context, and I contend that this move creates problems for Badiou's account.

Limitations and Liabilities in Badiou's Meta-ontological Subtraction

Meta-ontology is the discursive framework in Badiou's project to handle typically *philosophical* problems and concerns. For example, the first meta-ontological move of his ontology—the decision—is occasioned by the more common philosophical question concerning whether being is *One* or *many*; I discussed this decision and its implications in Chapter 1. As a result of the decision, Badiou develops a vocabulary for handling this question according to the subsequent claims that *ontology is a situation* and *being is multiple*. These are the meta-ontological means of cashing out the strictly mathematical ontology to come. A second meta-ontological move is to prohibit approaches that would claim ontology *is not* a situation. Following the decision in *Being and Event*, Badiou enacts a *restriction* against so-called "philosophical 'ontologies'" that turn ontology toward an object of inquiry more-or-less alien to the structures used to investigate it (on the one hand), or the nihilism of the Heideggerian position (on the other), which replaces philosophy with poetry and subsequently claims the enterprise of metaphysics is at its end.

For example, Heidegger (2000, 15–16) collapses the distinction between Being and becoming, a division—or restriction—he argues is invoked by Plato and Aristotle. These so-called "restrictions of being" are understood as different terms that are set up against the concept of being. *Becoming* is typically set against Being to explicate the problem of change; *seeming* marks the problem of appearances; *thinking* is used to illustrate the difference between the mind and the world. Heidegger sees these distinctions as false ones and takes as his task the demonstration of a "fixed continuity," or the shared meaning of becoming, seeming, and thinking with *physis*. In Heidegger's case, the restriction is artificial since it elevates *nomos* over *physis*; as such, the only way of dealing with *physis* is in the fragments of Heraclitus or the poetry of Parmenides. Indeed, the three reasons Badiou offers in *Being and Event* for rejecting "philosophical ontologies" may also apply to Heidegger's position. The 'experience' of *physis*, while it may be somehow intelligible it is not *expressible* in precise terms; as such, the recourse to the poetic is a failure of degree, both with respect to language and experience.

However, Badiou explains that a poetic ontology "finds itself in an impasse of an excess of presence, one in which being conceals itself" (*BE* 10). This is a failure of ***kind***, and one occurring at the conceptual level. Because the ontological difference is too radical and too comfortable with the prospect of the end of metaphysics, it goes beyond the tolerances of an investigation—ontological or philosophical—into being qua being. Badiou's solution, then, is to remove the extremes reflected in a Heideggerian approach. Restricting ontology against philosophy and poetry depends requires an object that can be discovered by inquiry, that is conceptually familiar, and linguistically expressible. This object, the void, is discovered subtractively and conceptually familiar on account of both its philosophical valence and mathematical significance; it is linguistically expressible insofar as the language put to it is that of mathematics. Badiou thus restricts philosophy to the specific operation and evaluation of truths in its conditions (art, science, politics, and love). A third move is accomplished by subtraction as a disposition as an imperative to continue the analysis of any situation toward the multiplicity that forms its foundation. Together with the restriction, subtraction forecloses against philosophical tools when it comes to investigating multiple-being, and mathematized ontology is elevated as the tool best suited to this task.

At first, Badiou's work in the restriction seems standard; positioning his work against Heidegger's seems fairly obvious, since it is Heidegger who revives the "being-question" with which Badiou's ontological project is primarily concerned. However, his opposition to Heidegger on this score requires that Badiou be able to situate his commitments in a philosophical (read: meta-ontological) register, both according to the philosophical problem identified in the decision and in relation to competing philosophical proposals on this score. The opposition to Deleuze is quite different, since—and as I have endeavored to demonstrate in the preceding chapters—their common commitments to multiplicity offer them something of an equivalent language in which to speak. In responding to broadly philosophical or meta-ontological demands, ones where common language and disposition aren't readily available or can't be constructed, Badiou abandons his ontological proposal and the resources it offers to stand as a genuine alternative in this broader conversation. In "The Question of Being Today," the essay opening *Briefings on Existence*, the decision against the One is the same; however, instead of inconsistent multiplicity at the end of ontology, Badiou posits an indeterminate, undefined *nothing*. In abandoning the precise sense of inconsistent multiplicity as no-thing, both a limit and a liability emerge in Badiou's meta-ontological analysis.

Just as *Being and Event* opened with the decision and an assessment of its implications and consequences, so *Briefings on Existence* opens with a refreshed take on the most fundamental position of Badiou's ontology. He justifies the position that being qua being is "the radical manifold or a multiple that is not under the power of the One" (*CT* 35). The first of three subtractive commitments Badiou unfolds is consistent with this claim. He says, "pure multiplicity, or the manifold unfolding of the unlimited reserve of being as a subtraction from the power of the One, cannot be consistent on its own" (*CT* 35). This is an assertion of Badiou's basic commitments—being qua being is multiple, not One; multiple being is pure multiplicity—but it includes a new component: subtraction is the active principle *resisting* the One. Because pure multiplicity is inert, it cannot animate itself; the "manifold unfolding the unlimited reserve of being" is subtraction's work. On this description, it is not clear what Badiou means by subtraction, except that "manifold unfolding is not constrained by the immanence of a limit" (*CT* 35–36). Subtraction is perhaps, in this context, the mechanism for pursuing and securing pure multiplicity past any limitation that might emerge. While ontology depends on the boundary present in the void set to begin its operation, Badiou describes here a procedure that exceeds limits to arrive at pure multiplicity.

Badiou takes this a step further by claiming that the subtractive disposition, when "complete," will discover "a multiple of nothing. The subtractive also amounts to the following: instead of conceding that for lack of multiple there is the One, assert that for lack of multiple there is nothing" (*CT* 37). Badiou's assertion of the identity of multiple and nothing is familiar insofar as it is introduced in Meditations 4 and 5 of *Being and Event*. There, the ascription of nothing to inconsistent multiplicity is to reinforce it as a site where existence is prohibited. In the ontological context (as opposed to the meta-ontological context), there are a series of positive indicators to keep nothing from slipping into a kind of nihilism. The status of nothing as inconsistent multiple and as the precedent of any consistent multiple all serve to keep a "productive nothing" in view. Put another way, the specifications of multiple as inconsistent, and then as nothing, are appropriate if they are specifications of an idea with a conceptual anchor as many qua multiple is. However, at the meta-ontological level, these specifications are removed entirely.

In the third subtractive commitment, Badiou restricts both definitions of the multiple (*CT* 37) and its naming as multiplicity or multiple (*CT* 38). To submit "multiplicity" to this naming, Badiou claims, is "to already enter into what Heidegger calls the 'process of limiting Being' by appealing to a delimiting norm. And the One would return" (*CT* 38). To define the multiple, then, amounts to an assertion of the One over it; this results in nominating or naming being with something other than the void Ø, which *in its precision* reinforces the inconsistency of inconsistent multiplicity. Indeed, general subtraction applied *after* the specification of multiple as inconsistent multiplicity and the assertion of Ø as its proper name is acceptable, since the limits of ontological operation are established; thought can compensate for a lack of access beyond the void set because it can form that void set into one. However, when subtraction finds being is *nothing*, there is a transformation of the problem on which the decision is founded. In other words, in trying to outrun Heidegger, Badiou insists on an indefinite nothing, which is neither One, nor multiple bearing the robust resources of his mathematical apparatus. In his desire to protect being qua being against the inappropriate control a definition would exert over being, Badiou's walking back from calling it multiplicity or inconsistent multiplicity draws his account perilously close to a kind of ineffability, thus *weakening* the protection he claimed to have installed against ontotheology. While Badiou does not posit an absolute or a god, the reticence to name or mark the foundation of being qua being opens the projects to objections of venerating "that which cannot be named."

In "The Question of Being Today," Badiou's goal is to carry the subtractive impulse that serves his ontology well enough to a broader philosophical position. To oppose his program Heidegger's, Badiou chooses the meta-ontological register, where subtraction brings thought to nothing without the tools to build back or out of that nothing. Badiou offers a general orienting remark when he says, "an axiomatic thought seizes upon the disposition of undefined terms" (*CT* 38). By this, Badiou seems to be claiming that set-theoretic ontology is suited to the program he describes here, since it is a program that depends on undefined terms for its functioning. However, the void set and its proper name Ø, while *empty* is hardly undefined. Indeed, the specific mathematical and ontological function of the void set *requires*, as Badiou demonstrates in Meditations 4 and 5 of *Being and Event*, a proper name in order to serve as the site of ontological initiation. But Badiou vacates this possibility when he claims that his solution to the problem of One and many escapes definition and limitation. In his global, meta-ontological subtraction, Badiou removes the anchor by which thought—axiomatic or otherwise—might be able to cope with multiplicity or multiplicity-as-nothing.

Badiou's entanglement with Heidegger in "The Question of Being Today" reveals a kind of limitation to the application of what I've called a general subtraction; when it is deployed in the meta-ontological register, it is emptied of the features that tie it to key ontological concepts, emptied of the things that give Badiou's ontology its distinctive character and the teeth of the restriction. A similar liability attends Badiou's engagement with Kant; in this case, Badiou's interpretation of the transcendental object = *X* and the transcendental unity of apperception as forms of void "withdrawn from being" (*TW* 143) creates problems for Badiou's own account of the void, even in its ontological richness.

In Chapter 1, I described two broad engagements with the manifold in Kant's *Critique of Pure Reason* The first, in the Transcendental Aesthetic, seeks to transform a merely passive engagement with multiplicity into an active empirical intuition, the material cognition requires to begin its work. The analysis of empirical intuition is constituted in relation to an appearance, intuition's "undetermined object" (Kant 1998, A20/B34). As the source of these undetermined objects, the manifold presents an array of data to which the faculty of sensibility—that faculty responsible for the production of intuitions—responds. Kant's claims regarding the nature of an appearance in the Transcendental Aesthetic allowed me to conclude that "the being of the manifold" is nothing to the mind other than undetermined. As undetermined, the manifold does not demonstrate its own organization; a "structuring intellect" arranges the appearance of the manifold from without.

The second encounter with the manifold is found in the Transcendental Deduction. I focused primarily on the A Deduction and its introduction of the transcendental object = *X*. I explained that the object = *X* functions as a placeholder necessary to secure the continuity of experience, and like the manifold of the Transcendental Aesthetic, the object = *X* is undetermined with respect to its content. It takes as its correlate the transcendental unity of apperception, which functions as the general condition by which cognition is possible. From the

B Deduction, I emphasized the work of combination (*Verbindung*) in connection with representation; representation provides the 'material' for combination and indicates the limit of cognition's reach. In other words, cognition can only operate on representations of an object, never the object itself.

Johnston (2008, 352–353) insists that Badiou's work remains in the grip of a Kantian structure in the form of the count-as-one, which he posits as an active mediation between inconsistent and consistent multiples. Further, the discussion of the inconsistent multiple (particularly as Badiou addresses it in Meditations 2–4 of *Being and Event*) installs inconsistent multiplicity as a version of a "thinkable but unknowable" noumenon (Johnston 2008, 353). Johnston, as well as Pfeifer (2015), have each raised concerns about the traces of Kantian structures in Badiou's work, in spite of Badiou's claims that, as Johnston (2008, 348) points out, "Badiou balks at Kant's invocation of the ostensible 'limits of possible experience,' insofar as this boundary line partitioning noumena from phenomena entails the prohibition of constructing a rational ontology."[3] The main objection Badiou raises against Kant, then, is that the distinction of phenomena and noumena maintains the impossibility of a *rational* ontology, and may only posit something like an ontology of presence, requiring access "beyond the noumenal veil" for an assessment or exposure to being.[4]

Questions—like those Johnston and Pfeifer raise—concerning *who counts*? and *where from*? are valuable, though are perhaps less devastating for Badiou's account when it is clear that ontology is a mechanism for mapping the relationship of structure to its suture to being at the void point; ontology is not, as I have attempted to show in my account of general subtraction, *prospective* but *retrospective*. This deflationary strategy, while perhaps less satisfactory than an ontology that generates objects using count-procedures (a special ontology) is in line with the restriction of ontology from the dynamic work of philosophy in Badiou's work.

I maintain a concern about Badiou's relation to Kant, however, over the notion of the void itself. In "Kant's Subtractive Ontology," Badiou reads the transcendental project through the lens of his own ontology. In particular, Badiou seizes the transcendental object = X and the transcendental unity of apperception as examples of Kant's "subtractive radicality." Badiou says,

> The feature common to both originary apperception as transcendental proto-subject and this *x* as transcendental proto-object is that, as the primitive invariant forms required for the possibility of representation, this subject and this object remain absolutely un-presented: they are referred to, over and above all possible experience, only as the void withdrawn from being, for which all we have are names. (*TW* 142–143)

Badiou equates the transcendental object and the transcendental unity of apperception as the foundation for possible experience, both empty of content and absolutely necessary for thought to proceed. Badiou's insistence that these structures are un-presented situates them analogously to the void in his own work.

The procedure of general subtraction is charged with discovering the limit of thought; in Badiou's work, the general subtractive procedure ends in the discovery of the void, which functions as the presentable feature of inconsistent multiplicity. Above, Badiou suggests the object = *X* serves a similar function in Kant's work; it names the boundary and stands-in for what thought cannot grasp. Badiou equates the object = *X* and the void further by describing the object = *X* as "an object that is nothing," a claim I discussed in connection with the presentation of the void set in Chapter 3 (*TW* 142). Badiou finds the Kantian attempt admirable but retreats to a critique of Kant over his requiring both the transcendental object = *X* and the transcendental unity of apperception. These 'two voids,' Badiou claims, require a *relation*, so Kant's work cannot constitute a fully subtractive ontology. In other words, subtraction must resolve to a *single* void, and not two.

I propose that Badiou's reading of Kant has problematic implications for Badiou's own project, which become clear on inspection of the transcendental object = *X*. The transcendental object = *X* is required for thought in the same way that the void and its esoteric mark are necessary for the determination of consistency; in fact, Badiou suggests these two elements are the same. Additionally, the content of both the object = *X* and the void are the same for Badiou, insofar as they both consist of nothing (no-thing). This opens a question, then, about how these respective void-markers, discovered by subtractive procedure, maintain their status as no-thing. However, the identification of transcendental unity of apperception as a void offers some correlate support for keeping the transcendental object = *X* legible and purposive. Together, these commitments shed light on the uncomfortable consequence of Badiou's aligning the void with the transcendental object = *X*; minimally, the coincidence of Badiou's project with Kant's here requires the deployment of an analogous, non-subtractive mechanism to keep the void a void—a kind of faith, perhaps, that maintains the presence of the void as such at the boundary of inconsistent multiplicity. This is another way of conceiving of the noumenon in Badiou's work, though from an inquiry into the void and not, as Žižek (2004) has argued for Badiou's dependence on a gap between the noumenon of the Real and the phenomenon of any world.

It may be that a deflationary account of ontology in Badiou's work, one that emphasizes its status as static and as a kind of mapping technology (as I proposed in the preceding chapter) each allay the worries about Badiou's smuggling in an active principle in the form of the count, or his failure to adequately bridge ontology and phenomenology. For my part, the analysis of the void alongside its Kantian correlates requires Badiou's reader to maintain the presence of inconsistent multiplicity beyond the void. This gesture—an honorific to Cantor's innovations—is procedurally overwritten in the presence of the void Ø. Further, this charge deposits Badiou's project among ontologies of presence; counterintuitively, perhaps, the presence in Badiou's ontology is an absence.

Badiou's objections to Kant's transcendental object register divergences in the accounts of local subtraction with which I open this chapter. Badiou could quite easily extend his criticism of Kant's project against Deleuze's; in other words, a truly subtractive ontology dissolves *all* bonds in order to reach no-thing and ontology of relation still depends on one thing. I am reminded of Peyrol's objection to subtraction in "The Fascism of the Potato," in which Peyrol objects that the subtractive disposition Deleuze and Guattari offer requires *both* one and multiple, the two (*FP* 199). Peyrol's requirements are to seek the condition of transformation, the "one of the multiple" to rupture the existing order. Badiou's critique of Kant in "Kant's Subtractive Ontology," and in the procedure of general subtraction I've identified above, offers a means of identifying that rupturing condition, minimally an identification of (and maximally a 'return to') the no-thing of the void point present in any situation.

Summary

The claim Badiou stakes regarding any comparison between his program and Deleuze's is that the central problem with which philosophy today is concerned is an "immanent conceptualization of multiplicity" (*DCB* 4) to be discerned in their respective approaches. As perhaps the most sustained of Badiou's combative texts, *Deleuze: The Clamor of Being* makes a number of attacks on Deleuze's program; chief among these is the claim that Deleuze is ultimately committed to a form of transcendence in disguise, implying that only Badiou's approach maintains a grip on multiple without collapsing into transcendence. An analysis of subtraction suggests otherwise, since in both Badiou's *and* Deleuze's programs some anchor of multiplicity is accessible to thought, the commitments of whatever version of multiplicity that system depends on are expressive within that system.

In Chapter 4, I discussed the way that Deleuze's virtual idea is a relation between differential elements resulting in pre-individual singularities. The connections holding among elements and their relations are themselves a genetic process. Subtraction offers an "immediate" access to multiplicity, insofar as the suspension of the unique instance brings inquiry to multiplicity. I demonstrated this in the analysis of the diagram and explication attending Deleuze and Guattari's "Example 1" in the previous chapter. There, the language of Riemannian multiplicity and principles of construction appropriate to continuous multiplicity are readily available. In Badiou's program, the general subtraction I described above shows that every consolidated unity bears the mark of inconsistent multiplicity; as such, multiplicity is immanent and *latent* in Badiou's account. The void point is buried and can only be uncovered with the requisite subtraction to observe situation as constructed according to principles of the consistent multiple, and to subsequently demonstrate the link of these consistent multiples to their foundation in the void set. It is from this vantage of a commitment in common, that of subtraction and its work in demonstrating the link to multiplicity—whether it is the dynamic, genetic multiplicity Deleuze and Guattari proffer, or the static void point forming a limit for thought in Badiou's work—that I now return to the engagements I outlined Chapter 2. The resources developed through a thorough investigation of structure and procedure afford a clarity around the consistent critiques each party brings against the other.

Conversation and Critique

In this section, I address the conversation I constructed in Chapter 2. There, I proceeded chronologically, beginning with "Rhizome" in 1976 and continuing to the publication of *Deleuze: The Clamor of Being* in 1997. With the benefit of the resources I developed in Chapters 3 and 4, I focus on and expand two points of discontinuity or divergence between Deleuze with Guattari and Badiou. First, and using the apparent mismatch in types of multiplicity prescribed in "Rhizome" and "The Fascism of the Potato," I return to critiques surrounding the *structure* of multiplicity raised across the 20-year period I've identified. Second, these critiques surrounding structure offer an occasion to reflect on the stasis and rigidity with which Badiou's ontological address of multiplicity is presented, a critique broadly aimed at *procedure*. Third and finally, I address Badiou's specific criticism of the presence of "monotonous productions" in Deleuze's work, a critique which offers an opportunity to consider the way mathematics is supposed to function in their respective work.

A Mismatched Multiplicity

In "Rhizome," Deleuze and Guattari take Maoism to task over the dictum "one becomes two" (*R* 51). For his part, Peyrol takes this difference in phrasing to be significant enough to lob an ad hominem at Deleuze and Guattari, insisting that "only a moron" would get this idea wrong. The presentation Peyrol offers as a replacement—"one divides into two" (*FP* 194)—is clarified by recognizing that scission expresses a dialectical matrix of division and not of alienation. If the latter, tracing the movement of the concept, then attending to the dictum as a scheme of becoming is useful; but in the lectures opening *Theory of the Subject*, Badiou is clear that scission follows the former. In the approach Badiou prescribes there, following his work with Hegel, temporary unities are divided according to their component parts and the relations between these; the relation of contradiction is revealed through something essentially "returned to itself" from another place and signified in the formula $A = AA_p$. Badiou explains that the bourgeois world facilitates a contradictory relation of bourgeois and proletariat, insofar as one only sees the other against the background of this world. Scission is a kind of pause and recollection, suspending—for the sake of analysis—alienation, the movement 'from' a dialectical position initiated by contradiction.

In Peyrol's "The Fascism of the Potato," scission takes up this task, emphasizing a temporary stasis, to fix the "one of the multiple" capable of revising the mass and its direction. Deleuze and Guattari seem to invert the Maoist image by insisting on images of root and tree; Peyrol suggests, by contrast, that the one is already and immediately two, already divided, and will return to that position after a temporary one is eclipsed. Peyrol insists the "one of the multiple" in May '68 is class, but he notes this only after acknowledging the diversity of the mass in his enumeration of the groups comprising "horizontal storm" (cf. *FP* 192). In Chapter 4, I explained that the function of scission is, for Peyrol, essentially revealing to each of the groups the background against which they are organized; in identifying this repressive background, it would be possible for each of these groups to band together and successfully re-write their present by revolution.

The tangible or visible nature of this particular mass is useful as a fulcrum, both for understanding Peyrol's critique of Deleuze and Guattari and Badiou's move away from materialism as the central paradigm around which his work is organized. As I pointed out in Chapter 2, Peyrol's objection to Deleuze and Guattari's "Rhizome" is a limit imposed by its avowed 'horizontality.' In this flattening, Peyrol argues that it gives way to a kind of fascism, invoking Pinochet in his warning call to "sectarians of the Rhizome!" (*FP* 197). Deleuze and Guattari address fascism in their discussion of multiplicity in "Rhizome", describing it parenthetically as "a horrible multiplicity—defined by its lines or dimensions precisely laid out on a plane of consistence" (*R* 55). Its horror consists not in its outcome, but in its mode of ***organization***; its precision and order forecloses against a creative tangle and particular results issuing from this tangle, "lines of flight" from the mass of connection. The horror is in stasis, in superimposed organization, or—worse—*supra*-imposed organization designed to straighten out a dynamic multiplicity otherwise operating according to an internal logic.

In their discussion of a rhizome's fourth principle, that of "asignifying rupture," Deleuze and Guattari explain that the movement interior to the rhizome must "be relative, always intertwining, the one caught up in the other (*R* 55). As such, and in the context of May '68s "horizontal storm," it is not a mere enumeration or identification that Deleuze and Guattari champion. Rather, it is the movements among these *in their particularity and in their present*, relative to one another, that interest Deleuze and Guattari. This commitment appears in Deleuze's account of the social Idea in *Difference and Repetition*, again in the analysis of the war machine and a-centered systems in "Rhizome", and in the discussion of the cogito in the first example in *What is Philosophy*? The transformation these transactions occasion is not necessarily revolution, but re-formation. Even if May '68 was successful on the terms Peyrol prescribes, it may be the case that Deleuze and Guattari would have been surprised by it, but it is also possible that they would be indifferent to it qua expression and more interested in the dynamic and particular connections giving rise to it.

From this vantage, it might be reasonable to dismiss Peyrol's critique of Deleuze and Guattari in this period as simply missing the point and move on. In his commitment to transformation and a constant attention to the signs permitting or prohibiting it, the Maoist Peyrol fails to see a bigger picture and project with which Deleuze and Guattari are concerned: thought in a much longer time frame (Deleuze and Guattari are "geophilosophers," after all) and an overturning of thought toward difference and disjunction, rather than conjunction and identity. However, and for the purposes of this study, it offers a moment's pause to consider these structures of scission and rhizome in tandem. It also offers a chance to consider again Badiou's ontological turn in light of the critique he raises in "The Fascism of the Potato." In Chapter 4, I explained the exhaustion Badiou observed with respect to materialist politics and the disposition of his theoretical work toward structures capable of discovering the irruption of the Real in any situation. Though there is a kind of displacement of Badiou's attention, Power and Toscano (2009), Bosteels (2012), and Dosse (2010) have each convincingly argued for the intractability of Badiou's ontology from its political roots. This link is reinforced by attending to the glancing comment Deleuze and Guattari make about fascism and its structure. Badiou's ontological turn is of course politically and transformationally oriented; however, it is in these early encounters that the seeds of Badiou's "ontology of the multiple," positioned against Deleuze's are planted (*DCB* 3).

These early encounters also offer resonances of the critiques to come. For example, the conceptual objections Deleuze and Guattari raise against Badiou's project in "Example 12" of *What is Philosophy*? have roots in these early critiques. Though logic, that broader operative context in which Deleuze and Guattari engaged Badiou's work, is not fascism, it is "motivated by a real hatred" and "rivalry with, or will to supplant philosophy" (*WP* 140). This is accomplished by first fixing a concept to a plane of reference and then reducing it to a rigid system of predication; Deleuze and Guattari object to transformations of structure that straightens the pathways of intensive ordinates to becoming lines in a proof. When logic does admit a philosophical concept, it casts the concept as "the subject's pure and simple opinions, of subjective evaluations or judgments of taste" (*WP* 141). The concept is admitted as empty because it lacks the rigor of number or "the well-ordered set" (*WP* 141), and blind since it cannot speak in the language of a doubly-fixed frame of reference.

Unlike a Russell, Frege, or (a caricature of) an analytic philosopher, Badiou has a peculiar response, stemming from a meta-ontological move required by his work. The radical thesis is presented as a hypothetical, and there are good reasons to take it seriously as such: *if* being qua being is a pure multiplicity, only *then* is mathematics ontology. The inaugurating *if* takes a strong position insofar as it shifts the being-question out of the domain of philosophy, where it is beset by so-called "ontologies of presence" (think Levinas, or Plato of the forms), or Heidegerrian abdications to poetry. These options are in fact *too* devoted to correspondence and are frustrated in the face of peculiarly human being. Instead, Badiou assigns to ontology the being of the pure multiple and that alone, the task, as Badiou (with Haéri 2016, 49) says, "of the thinking of what there is, of what is." As I have shown in the preceding, particularly regarding Badiou's commitment to the void, his concern is to effectively map patterns of existence using the tools of mathematics. Put in a more direct reply to the critique Deleuze and Guattari raise, Badiou uses logical tools but in his ontology, in particular, these are loosed from the requirements of logical correspondence insofar as they do not pronounce "true" or "false," and rather *describe* the structures of what is. This is another way of tracing the transformation from Badiou's earlier work (e.g., in *The Concept of Model*), where tools of mathematical logic are set to operation directly on political realities, to map out a transformative science capable of withstanding incursion by bourgeois ideology (Brassier 2005, Livingston 2011). Badiou's work may also be unmoored from a scientific frame of reference because of its founding commitment to being qua being as pure multiplicity.

It is in this way that Badiou's account is further distanced from a philosophical sense of "the many," since the content of inconsistent multiplicity is an infinite multiple that may not be closed or completed; it is not bodies and void in an atomist or Lucretian sense. In *Briefings on Existence*, Badiou claims Democritus is committed to a subtractive disposition, expressed as "the dismissal of the One," occurring in his work "through a dissemination and appeal to the void" (*CT* 35). He takes the subtractive disposition of Lucretius further, claiming that in *De Rerum Natura*, "Lucretius is the one who directly confronts thought to subtraction from the One, which is none other than inconsistent infinity, that is, what nothing can collect" (*CT* 35). Whether these assignations are correct is, of course, significant; it is possible to note, though, given the trouble I've pointed to above concerning Badiou's meta-ontological deployment of subtraction that the comparison is *at best* only one of loose analogy. Here, the terms of inconsistency and nothing float free of their rigorous mathematical mooring, which compromises the comparison to Lucretius or Badiou's own commitments, neither of which is desirable. The atomist position has, perhaps more in common with Peyrol, who is committed to an empirical expression of mass, its specific diversity, and its resistance to totalization.

Structures, Rigidity, and Stasis

In their response to Ricardo and David Nirenberg, A.J. Bartlett and Justin Clemens (2012, 368) counter the former pair's critique by explaining that a mathematical ontology, for Badiou, is "a figure of philosophical *fiction*." As the preceding work has demonstrated, mathematized ontology must essentially function as a "productive placeholder" so that philosophy may be liberated from questions of being and the traps attending these and instead investigate the truths stemming from transformation.

A precise statement of generalized subtraction reveals an ontological map of an event that, while neutral, is not an origin in the sense of a generation or production. The work of mathematized ontology is simply to map or describe the structure of that transformation; this is a consequence of Badiou's claims concerning scission and, when combined with the account of general subtraction I've developed above, it demonstrates that though ontology can build, it is also effective in *reading subtractively*. Using the tools of axiomatic set theory and the relations of belonging and inclusion, reveals a maximum of three types of ontological situations: normal (what belongs is included), excrescent (what is included does not belong), and singular (what belongs is not included). The structural features of these three types of situation are revealed in the general subtractive disposition, which divests procedure to discover the void point at the foundation of any ontological structure; this also has the effect of showing the limit of thought, insofar as thought consists in the tools of axiomatic set theory. The positive apparatus of structure and procedure present in Badiou's ontology, though presented *first* in the system, offers a retrospective *mapping*—instead of asking "what went wrong in May '68?," Badiou's ontology can offer a descriptive framework of Pauline or Cantorian transformations that is also prescriptive insofar as other concrete candidates for transformation—Tahrir Square, for example—are structurally legible.

Rigidity also appears here in Deleuze and Guattari's objection to Badiou's use of the count, identifying "the set's subjection to a regime of the 'counting-as-one' (bodies or objects, units of the situation)" (*WP* 151). Though a function expresses relations among components circulating in the frame of reference (the site of scientific thought), the method of set construction using the count applies, by Deleuze and Guattari's lights, a functional relationship to freely-circulating elements in the plane of immanence (the site of philosophical thought). As I pointed out in Chapter 2, this yields an objection to Badiou's work that claims it is too rigid for philosophical operation. Any event, irrespective of the domain in which it occurs (art, science, politics, love), has the same ontological structure, apparent in the asymmetry of belonging and inclusion. This implication—that an event bears universal markers wherever it erupts—is problematic for Deleuze and Guattari, as a capitulation to identity or "the same". As the analysis of structure in Chapter 2 showed, the multiplicity they affirm as the condition for any empirical or conceptual expression is radically *particular*, determined according to the pathways developing among the nodes or elements themselves. The only "universal," if it can be identified as such, is the basic definition of multiplicity as between at least two elements. Deleuze and Guattari abandon any pretensions to prediction or prescription when it comes to a multiplicity's expression. Put another way, predictability is the price of creativity.

I've used the preceding pages to demonstrate the ways a more thorough account of multiplicity deepens the critique and conversation persisting between Badiou and Deleuze. When Deleuze and Guattari engage traditional materialism, Maoism, or Badiou's ontology, the tendency is to read its dependence on models of reproduction from a single, *genetic* point; Deleuze and Guattari superimpose the requirements of multiplicity as they conceive it to reveal limitations in the old way of thinking. Badiou's account of scission, when read alongside an articulation of general subtraction, shows no such commitments. As my analysis demonstrated, this genesis is no part of the ontology developed in *Being and Event*, since the inconsistent multiple and void set are inert; procedure in the form of Zermelo–Fraenkel axioms is shown to compose something from the nothing of the void. For its part, the count and its functioning are—just as the nature of any one is shown to be multiple—revealed or discovered retrospectively; the presence of a situation indicates the count has been applied. This recognition requires Badiou's ontology to operate in a more-or-less *hypothetical* frame of reference. Evidence for this is available in Badiou's preface to *Being and Event*, and particularly Part IV, which I discussed in Chapter 4. In this hypothetical frame, Badiou's ontology offers tools of mapping in terms congenial to the antecedent claim of the "radical thesis," the assumption that being qua being is multiple. Additionally, Deleuze and Guattari's persisting—and compelling—critique is that the schematic prescribed both in traditional forms of materialism and the project of Badiou's ontology attempts to develop the new or novel from rigid, universal frameworks. In response, Deleuze and Guattari reassert their emphasis on multiplicity as *between* elements, pathways forged and re-forged in local bundles of connection. This interior movement—striving for and maintaining horizontality, subtracting the one to make the multiple—produces transformation or expression on a smaller, more frequent, and potentially more legible scale.

It is in examining this commitment to one-ing, counting, or isolating a class condition that Badiou, writing both as Peyrol and himself, raises the critique of stasis against Deleuze and Guattari, identifying as problematic any multiplicity that in its practice of mere affirmation has no transformative traction. An examination of Deleuze and Guattari's own response to claims of fascism or problematic inaction directs the reader's attention to the structure of the multiple itself. Deleuze and Guattari assign fascism to multiples that are essentially one-dimensional, with a single trajectory and in which no deviation is possible; in these cases, creative escape is foreclosed against. In light of this critique of rigidity, whether Deleuze and Guattari bring it against fascism or the misguided application of logical requirements to philosophical concepts, Badiou's project once again comes into view. I explained that though Badiou's work develops its ontology from a commitment to multiplicity, it is not a creative, dynamic, or self-governing multiplicity; rather, it is a multiplicity that trades in universal types, situations organized according to the combinations of belonging and inclusion. These types hold irrespective of the complexity that may develop in any world (the ontological apparatus Badiou develops in *Logics of Worlds*). Though these universal types are both descriptive and prescriptive, they do not support *creativity* per se; one must move to the emergence of the faithful subject in *Being and Event* to find the dynamic moment in Badiou's ontological apparatus.

Monotonous Productions, Multiplicity, and Model Horizons

In Chapter 2, I explained that Badiou charges Deleuze with "monotonous productions," claiming that Deleuze's work is concerned with "composing a very particular regime of emphasis or almost infinite repetition of a limited repertoire of concepts, as well as a virtuosic variation of names, under which what is thought remain essentially identical" (*DCB* 15). Evidence for Badiou's claim is available in the work I've done, for example, to develop Deleuze's and Deleuze and Guattari's commitments; overturned Platonism, the work in "Example 1," the analysis of the social Idea as a virtual Idea, the war machine, and the rhizome itself are all different means for presenting the structure and procedure associated with Deleuze's account of multiplicity. Though he claims this consistency is a virtue of Deleuze's account, Badiou sees the proliferation of examples as a matter of reproducing the conditions of thought at the cost of displacing the specificity and uniqueness of the example itself. The issue underwriting Badiou's criticism thus concerns whether Deleuze's framework admits of *too much* flexibility, such that its only trace or reliable characteristic is its having-been-enacted, the evidence of which is a series of productions that serve the form of thought, rather than its specific content.

I discussed in some detail the reply Daniel W. Smith ([2003] 2012) offers to Badiou's characterization of Deleuze's multiplicity as vitalist and nonmathematical. In short, Smith establishes that because Badiou is committed to only one pole of mathematical inquiry—the axiomatic—he fails to appreciate Deleuze's alignment with the problematic pole. On Smith's analysis, Badiou's selection of axiomatics and the assertion that mathematics = ontology each make it impossible to entertain any other possibility. The equivalence of mathematics = ontology only means a *particular kind of mathematics*, and as Smith ([2003] 2012, 288) explains, "the more precise equation however, would, be that 'ontology = axiomatic set theory,' since for Badiou it is only axiomatic set theory that mathematics adequately 'thinks' itself and constitutes a condition of philosophy." Indeed, it is clear that Badiou's commitment to axiomatic set theory structures his ontology according to rigid principles associated with demonstration, rather than the creative or constructive principles associated with problematics, at least as far as the ontology is concerned; it is also the case that the ontological principles associated with situations are explicitly set-theoretic, derived either from the Zermelo–Fraenkel system or the innovations by Cohen in support of the generic multiple.

In *Logics of Worlds*, by contrast, Badiou invokes complexity and density in the discussion of a topological space, "A mathematical concept that aims to rigorously think what is the 'place' of a being, its environs (or neighborhoods), its boundary, and so on" (Badiou 2009, 596). Here, principles of density, connection, and relation all offer a topological space its unique character; this, however, is the business of *appearing*, not *being*. Put another way, Badiou's broad mathematical paradigm does not exclude constructive or dynamic configurations, but these configurations are *ontologically* prohibited. Badiou's ontology, on subtractive analysis, yields just three types of situation, only one of which underpins transformation. The critique concerning monotonous production could be just as easily turned on Badiou's ontology, since it is governed by a restrictive vision of what can count for an ontological structure, and an equally restrictive vision of the conditions under which transformation occurs.

In "One, Multiple, Multiplicity," a text written in response to critiques of *Clamor* indicating Badiou's paltry attention to the Riemannian foundations of Deleuze's work, Badiou returns to mathematical resources (generally) and Riemann (specifically) to attend to their status in Deleuze's work. This short text begins with a series of affirmations of Deleuze's work, both in its practical resistance to the New Philosophers (*TW* 68) and its conceptual commitments to a metaphysics of multiplicity and singularity (*TW* 69). It includes a series of provocative engagements, some of which (as in Section 1 of the text) rehearse again Deleuze's apparent commitment to a metaphysics of the One-All and a vitalist multiplicity; the problematic commitment, discussed in Section 2, to the notion of multiplicity as a play between one and multiplicity, or *closed* and *open*; and in Section 4, an assessment of their respective readings of Spinoza and the status of analogy in Deleuze's work. Pertinent to the present discussion are elements of Section 3, in which Badiou discusses the deployment of mathematics and the explicit resources of the Riemannian paradigm.

Badiou affirms that mathematics has abstract and theoretical resources that far exceed empirical presentation. Mathematics, Badiou explains, "shows itself perfectly capable both of providing schemas adequate to experience, *and* of frustrating this experience by way of conceptual inventions that no intuition could ever accept" (*TW* 75). The instance Badiou begins with is the graze, a physical experience that can be conceptually accounted for in the tangent of any continuous curve; Badiou cites this on account of its similarity to a "superficial touch" or a caress in the empirical world, and its affinity to certain Deleuzian concepts (i.e., lines of flight). However, Badiou then points to a nineteenth-century discovery "that there exist continuous functions that cannot be derived *at any point*," to say that there are continuous functions for which a tangent cannot be identified. Badiou claims, "This is just a particular case of a general law: everywhere where mathematics is close to experience but follows its own movement, it discovers a 'pathological' case that absolutely challenges the initial intuition. Mathematics then establishes that this pathology is the rule, and that what can be intuited is only an exception" (*TW* 75). This is a revealing critique on Badiou's part. Of course, there are pathological instances in mathematics, and in this discussion Badiou raises a warning of sorts that one ought to not get overly invested in the commitments to differential calculus and the behavior of curves, and particularly as these prescribe anything about being. However, rather than analytical, this critique seems grossly prescriptive; choose the *right mathematics*, the one that doesn't give itself to pathological interpretations that might undermine the entire apparatus.

The resources of multiplicity (at least as these are associated with the differential so far) are less diverse, per Badiou, than "the variegated configurations proper to set-theory," which "provide an incomparably richer and more complex resource" (*TW* 76). On this point, Badiou cites the possible intervention of a generic subset in a partially-ordered set, claiming it "not only surpasses in violence, as a case for thought, any empirical rhizomatic schema whatsoever, but, by establishing the conditions for 'neutrality' in a multiple that is both dispersive and co-ordinated, it actually subsumed the ontology of these schemata" (*TW* 76). The subtext here is that the ontological rupture offered by the event is more of a rupture than anything the rhizome has on offer; the generic intervention into a situation re-orders the structure of the situation according to the violence of the void's irruption therein. Badiou then enumerates types of cardinality, including *inaccessible, compact, ineffable, measurable, enormous, Mahlo, Ramsey,* and

Rowbottom cardinals, that maintain specific principles for their respective constructions, each relying on apparatuses in excess of what's available in the Zermelo-Fraenkel axiom system. This move is authorized, at least in part, by the recognition late in *Being and Event* that the intervention of the generic multiple requires an expansion of set-theoretic principles beyond the initial constructions permitted under the Zermelo–Fraenkel paradigm. Badiou raises this set of examples to affirm the density of the concept of set construction, which functions in complex ways apart from the mere issue of number. Further, the principles for constructing any set and principles governing their content are diverse and are sufficient to generate the countable infinities—and kind of density—Badiou is after. In the critique that suffuses "One, Multiple, Multiplicities," Badiou doubles-down on the commitment to mathematics as ontology, insisting that the commitment to set-theoretic ontology follows its expansion to include robust and diverse forms of construction. These constructions exceed any form of complexity imagined by a rabbit warren, dandelion grass or any other form of rhizomatic development (though, somehow, they resist the esoteric or occult nature with which Badiou-as-Peyrol saddled Deleuze and Guattari's account of multiplicity in 1977).

A more faithful reading of Riemann, as Badiou explains in Section 3, sees Riemann's work as paving the way for the set-theoretic interventions of Cantor and Dedekind (*TW* 77). Badiou also insists that Riemann's considerations are mathematically Platonist through and through, requiring "a speculative framework entirely subtracted from the constraints of empirical intuition" (*TW* 77). On Badiou's assessment, Riemann is a poor fit for Deleuze's theory of multiplicity because Riemann's work, understood in an exact trajectory, expresses principles Deleuze himself would not accept. In my own discussion of Riemann's work in Chapter 3, I explained that he decamps to the conceptual in order to develop the multiply-extended manifold; Riemann is, in other words, invoking an abstract framework to invent the concepts necessary for his geometric intervention. Indeed, the *letter* of Riemann's work (per Badiou) is not only a kind of Platonism, but also neutralizes difference and locality insofar as the multiplicity Riemann prescribes "in effect can indifferently welcome numbers, points, functions, figures, or places, since it does not prescribe that of which it is composed" (*TW* 77). In failing to grasp the metaphysical commitments of Riemann's work, Badiou suggests that Deleuze in effect dooms the metaphysical paradigm of multiplicity invoked across his oeuvre.

This perplexity, borne from Badiou's commitment to an exactitude in the deployment of application of mathematical paradigms, may be relieved in part by attending to the function of mathematics in Deleuze's work. In his thoroughgoing investigation of Deleuze's use of mathematics, Simon Duffy proposes three general "components" to explain Deleuze's mathematic engagement. "The first component," Duffy (2013, 1) explains "can be characterized as the history of mathematics relevant to each of the programs or mathematical disciplines with which Deleuze engages, and the mathematical problems or problematics that are extracted from them." Readers of Deleuze are familiar with the more general identification of *royal science* and *nomadic science*, and Deleuze's typical selection of nomadic, or minor, paradigms according to which he articulates his system. This proclivity of Deleuze's underwrites the task of Duffy's text, which offers a historical accounting of the minor mathematical tradition on which Deleuze depends, in service of a third component, the "creation of new concepts by bringing together mathematical and philosophical problematics" (Duffy 2013, 3). However, it is in connection with a second component of Deleuze's use of mathematics, an effective transfer of the selected "interventions in the history of mathematics" to philosophical problems (Duffy 2013, 2) that a counter to Badiou's critique becomes available.

Duffy (2013, 2) says, "It is important to note that Deleuze eschews characterizing his redeployment of mathematical problems and problematics as simply analogical or metaphorical." Put another way, Deleuze does not play fast and loose with mathematical models, and similarly does not attempt to force "quantitative and exact" models into philosophical discourse, since these are the province of science and set the philosophical analysis on the tenuous ground. For example, one would not expect to find the results of a specific experiment in biology or chemistry, or a well-known problem from physics deployed as a model in Deleuze's work or in his collaborations with Guattari. Citing Deleuze's discussion in *Cinema II*, Duffy (2013, 3) explains, "Deleuze recognizes that citing mathematical notions of the exact kind outside of their particular sphere would rightly expose one to the criticism of 'arbitrary metaphor or of forced application'." This recognition at once reveals the *limitation* on mathematical application in Deleuze's work and the *liability* attending Badiou's own commitments; *forced application* is certainly the charge levied at Badiou by Deleuze and Guattari in "Example 12," and again in 2011 at Badiou by the Nirenbergs. The way out—a full restriction of ontology to mathematical outworking, to which Badiou is committed—will only limit the nature of the discourse altogether.

Duffy (2013, 2) points out that the mathematical models Deleuze deploys are models and "problems that are 'essentially inexact yet completely rigorous'." Quoting Deleuze (1995, 29) Duffy explains that a model's designation as inexact consists in its "conceptualizable character." For example, my presentation of the various instances in which Riemannian principles are expressed in Deleuze's work supports the recognition that Riemann's work is tapped precisely because it can be mobilized in diverse frameworks. Deleuze's multiplicity depends on the principles expressed in Riemann's descriptions of determinations and dimensionality; it could (and does) describe the basis for a differential geometry, but the principles of *n*-dimensionality and immanent measure are, for example, sufficiently inexact to be mobilized toward prescribing relationships and possibilities at the genetic foundation of a concrete case. I have endeavored to demonstrate this in several examples drawn from Deleuze's oeuvre.

Duffy's work offers greater insight into why Deleuze selects Riemann's model of multiplicity, rather than the exact model Badiou adopts, and it is, therefore, a better explanation for Deleuze's selection than only capitulating or falling back into Deleuze's preferences for the *minor* or *nomadic* sciences. Deleuze seeks a model with principles that can be exhausted in areas beyond their immediate or local disciplinary expression. Though Badiou insists that expanding the number of precise mathematical resources (in different forms of cardinality and their associated principles of construction) will ensure coverage of ontological diversity, Deleuze invokes a structure and procedure with 'maximal' expression. For example, in Plateau "1440: The Smooth and the Striated," Deleuze and Guattari investigate the structures and relation of smooth space to striated space in terms of mathematics, maritime parlance, textile creation and weaving, aesthetics (or "nomad art" [*TP*, 492–499]), music, and physics. They conclude the plateau with an injunction:

> Do not multiply models. We are well aware that there are many others … What interests us in operations of striation and smoothing are precisely the passages or combinations: how the forces at work within space continually striate it, and how in the course of its striation it develops other forces and emits new smooth spaces. (*TP* 499–500)

The value of the model, for Deleuze and Guattari, is found in the tools it offers for reading and mapping the particularity of the concrete case, the transaction *between* multiple and one constituting the substantive sense of multiplicity on which Deleuze—writing on his own and with Guattari—depend. In this way, an emphasis on locality or particularity realizes this "inexact yet completely rigorous" deployment of models. Badiou's identification of so-called "monotonous productions" is a failure (or a refusal) to understand the specificity of the conditions *producing* the example, the localized arrangement of abstract elements in a multiplicity linking up in a way that generates the expressions appropriate to it. The analysis, for example, of the difference between the diagram of Descartes' cogito in "Example 1" and its resonances with the "ox head wired for sound," that diagram associated with "Example 4" (both in *What is Philosophy?*) express this local difference; in the Riemannian analysis of the cogito, which I took up in the previous chapter, the structural features of the diagram are deployed not only in construction, but also as tools for *reading* the diagram itself, with attention to the particularity and specificity of its conditions.

Conclusion

I began this chapter by identifying a commitment in Badiou's ontological project to a general form of subtraction; this procedure serves as an injunction of sorts to read any consolidated situation—indeed, from any concrete instance—toward the identification of the void point and suture to being evident in any situation. I distinguished general subtraction from a so-called special form of subtraction, in which Badiou addresses the logical and ontological movement of truths emerging from any evental rupture. The presence of a general subtractive procedure identifies, along with a commitment to multiplicity, a second commitment in common with the Deleuzian program insofar as both deploy subtractive procedure to "make the multiple" in their respective work. An important difference obtains, of course, in that Deleuze and Guattari's subtractive procedure is a kind of normative requirement (which I established in Chapter 4), while Badiou's general subtraction is necessary only insofar as it serves that general commitment to seek the void point at the foundation of thought structured ontologically.

I showed that while Badiou's prescription for general subtraction offers a mapping strategy for the ontological makeup or framework of any situation built according to the robust components of his mathematized ontology, liabilities emerge when he deploys subtraction at a meta-ontological or philosophical level. In these cases, the tethers that maintain being qua being as *no-thing*—precisely the character of the inconsistent multiple, the proper name of being in the form of the void, and the relations of belonging and inclusion used to count the void set—are eliminated. This unmooring transforms Badiou's work into an ontology of presence, insofar as it installs an analogue that requires a certain kind of faith, but lacks the requisite tools for approaching its object. I demonstrated this by reading Badiou's meta-ontological subtraction in his critique of Heidegger in "The Question of Being Today" and his engagement with Kant in "Kant's Subtractive Ontology." These meta-ontological engagements identify a liability apparent in Badiou's work following from the strict assignation of ontology to mathematics.

I then returned to the conversation I opened in Chapter 2, addressing the critiques I enumerated there with the resources of more fully-formed understanding of multiplicity as structure and procedures for one-production in Deleuze's and Badiou's respective work. In the direct exchanges in 1976, 1977, and again in 1991, I showed that there are two major issues underpinning these critiques, issues which add dimension to the response of one program to the other. Deleuze and Guattari fail to appreciate Badiou's work from a procedural standpoint; that is, Badiou's ontology does not *generate* but rather functions retrospectively to map the structure of a situation, and similarly, *scission* does not generate but rather engages a kind of stoppage so that a transformative condition might be drawn from the background. A second issue is linked to structure, since the transformative condition can only emerge in Badiou's work from a *single kind of structure*, the singular situation. I used this insight to return to Badiou's critique of Deleuze in *Deleuze: The Clamor of Being* concerning what Badiou identifies as "monotonous productions" in Deleuze's work. I used this as an opportunity to reflect on the demands Badiou places on mathematics and, in reading "One, Multiple, Multiplicities," for example, the way these demands translate to his critique of Deleuze on exactly this point. I addressed Simon Duffy's careful account of the "inexact yet rigorous" functioning of the Riemannian model in Deleuze's work as a way of assessing Badiou's critique.

My intention in this chapter has not been to close a conversation; indeed, and if anything, the concluding remarks concerning models or the gesture toward Platonism in Badiou's arguments in "One, Multiple,

Multiplicities" ensure sites at which an informed assessment of multiplicity as structure and mathematics as procedure will deepen these portions of the discussion. Though I have not discussed the specific doctrine of the event, the specter of transformation haunts my own analysis, as it forms the practical and practicable expression of any ontological structure, any prescriptive procedure; I have shown in these remarks that its intuitive presence dictates the way structures and procedures are construed, and explicit engagement with a doctrine of event, as commitment-in-common and thus a center for conversation, forms the horizon of any future inquiry.

Notes

1. *Mathematics of the Transcendental* accounts for the complicated mathematical apparatus Badiou deploys in *Logics of Worlds*, one that serves to bridge the ontology and theory of being that occupies *Being and Event* with the ontology and theory of appearing *Logics of Worlds* is tasked with.
2. Paul Livingston (2012, 255/ff.) identifies the theory of ontological rank as a point of convergence between Badiou's "generic orientation of thought" and precedent, "criteriological" positions. Livingston's discussion of this aspect of Badiou's work spurred my thinking on this point.
3. In fact, Johnston identifies "three fundamental reasons why Kant functions as one of the main nemeses of Badiouian philosophy" (347); Badiou's rejection of the noumenal/phenomenal distinction is the second of these three. The other two consist in Kant's emphasis on subjective finitude, which ties philosophy and its prospects to the problems of human subjectivity (347), and finally Badiou's commitment to materialism, on the basis of which Badiou insists that his project is fundamentally incompatible with Kantian idealism (348).
4. Michael Olson (2009, 154) has suggested that an ontological reading of Kant's critical project transforms its central task according to "the subjective constitution of the objectivity of objects."

References

Badiou, Alain. [1998] 2004. "Kant's Subtractive Ontology." In *Theoretical Writings*, edited and translated by Ray Brassier and Alberto Toscano, 137–144. London: Continuum. (Cited in text as *TW*)

———. [2000] 2004. "One, Multiple, Multiplicities." In *Theoretical Writings*, edited and translated by Ray Brassier and Alberto Toscano, 67–80. London:

Continuum. Originally "Un, multiple, multiplicite(s)." In *Multitudes* 1 (2000): 195–211. (Cited in text as *TW*).

———. 2005. *Being and Event*. Translated by Oliver Feltham. London: Continuum. Originally *L'être et l'événement*. Paris: Éditions du Seuil, 1988. (Cited in text as *BE*)

———. 2008. *Conditions*. Translated by Steven Corcoran. London: Continuum.

———. 2009. *Logics of Worlds: Being and Event II*. Translated by Alberto Toscano. London: Bloomsbury. Originally *Logiques des mondes*. Paris: Éditions du Seuil, 2006.

———. 2012. "The Fascism of the Potato." In *The Adventure of French Philosophy*, edited and translated by Bruno Bosteels, 191–201. London: Verso. Originally Georges Peyrol. "Le fascisme de la pomme de terre." In *La situation actuelle sur le front de la philosophie*, 42–52. Paris: François Maspero, 1977. (Cited in text as *FP*)

Badiou, Alain, and Gilles Haéri. 2016. *In Praise of Mathematics*. Translated by Susan Spitzer. London: Polity. Originally *Eloge des mathématiques*. Paris: Flammarion, 2015.

Bartlett, A.J., and Alex Ling. 2014. "Translators' Introduction: The Categorial Imperative." In Alain Badiou, *Mathematics of the Transcendental*, edited and translated by A.J. Bartlett and Alex Ling, 1–10. London: Bloomsbury.

Bartlett, A.J., and Justin Clemens. 2012. "II: Neither/Nor." *Critical Inquiry* 38 (2): 365–380. https://doi.org/10.1086/662747. Accessed 1 September 2013.

Bosteels, Bruno. 2012. "Translator's Introduction." In Alain Badiou, *The Adventure of French Philosophy*. Translated and edited by Bruno Bosteels, vii–lxiii. London: Verso.

Brassier, Ray. 2005 "Badiou's Materialist Epistemology of Mathematics." *Angelaki* 10 (2), 135–150.

———. 1995. *Negotiations: 1972–1990*. Translated by Martin Joughin. New York: Columbia University Press. Originally *Pourparlers*. Paris: Les Éditions de Minuit, 1990.

Deleuze, Gilles, and Félix Guattari. 1981. "Rhizome—Introduction." Translated by Paul Patton. *I & C* 8 (Spring): 49–71. Originally *Rhizome: Introduction*. Paris: Les Éditions de Minuit, 1976. (Cited in text as *R*)

———. 1987. *A Thousand Plateaus*. Translated by Brian Massumi. Minneapolis, MN: University of Minnesota Press. Originally *Mille Plateaux*. Paris: Les Éditions de Minuit, 1980. (Cited in text as *TP*)

———. 1994. *What is Philosophy?* Translated by Hugh Tomlinson and Graham Burchell. New York: Columbia University Press. Originally *Qu'est-ce que la philosophie?* Paris: Les Éditions de Minuit, 1991. (Cited in text as *WP*)

Dosse, François. 2010. *Gilles Deleuze and Félix Guattari: Intersecting Lives*. Translated by Deborah Glassman. New York: Columbia University Press.

Originally *Gilles Deleuze et Félix Guattari, Biographie croisée*. Paris: Éditions la Decouverte, 2007.

Duffy, Simon. 2013. *Deleuze and the History of Mathematics: In Defense of the "New"*. Bloomsbury Studies in Continental Philosophy Series. London: Bloomsbury.

Hair, Lindsey. 2006. "Ontology and Appearing: Documentary Realism as a Mathematical Thought." In *The Praxis of Alain Badiou*, edited by Paul Ashton, A.J. Bartlett, and Justin Clemens, 265–290. Melbourne: re.press.

Heidegger, Martin. 2000. *Introduction to Metaphysics*, 1st ed. Translated by Gregory Fried and Richard Polt. New Haven, CT: Yale University Press.

Johnston, Adrian. 2008. "Phantom of Consistency: Alain Badiou and Kantian Transcendental Idealism." *Continental Philosophy Review* 41: 345–366. https://doi.org/10.1007/s11007-008-9086-5. Accessed 24 July 2015.

Kant, Immanuel. 1998. *Critique of Pure Reason*. Translated by Paul Guyer and Allen W. Wood. Cambridge, UK: Cambridge University Press.

Livingston, Paul. 2011. "Badiou, Mathematics, and Model Theory." Unpublished. Available online: https://www.unm.edu/~pmliving/Badiou,%20Mathematics,%20and%20Model%20Theory.pdf. Accessed 18 October 2011.

Livingston, Paul. 2012. *The Politics of Logic*. London: Routledge.

Olson, Michael. 2009. "Transcendental Idealism, Deleuze and Guattari, and the Metaphysics of Objects." In *Thinking Between Deleuze and Kant: A Strange Encounter*, edited by Edward Willatt and Matt Lee, 151–170. London: Continuum.

Pfeifer, Geoffrey. 2015. *The New Materialism: Althusser, Badiou and Žižek*. London: Routledge.

Power, Nina, and Alberto Toscano. 2009. "Alain Badiou and the Enemies of May." *Boundary 2* 36 (1): 27–46. https://doi.org/10.1215/01903659-200-022. Accessed 3 April 2014.

Smith, Daniel W. [2003] 2012. "Mathematics and the Theory of Multiplicities: Badiou and Deleuze Revisited." In *Essays on Deleuze*, 287–311. Edinburgh: Edinburgh University Press.

Žižek, Slavoj. 2004. "From Purification to Subtraction: Badiou and the Real." In *Think Again: Alain Badiou and the Future of Philosophy*, edited by Peter Hallward, 165–181. London: Continuum.

CHAPTER 6

Conclusion: Multiplicity, Ontology, Deleuze, Badiou

This book has been concerned with the (re)construction of a conversation between Badiou and Deleuze with Guattari in the period 1976–1997, identifying their major presentations of multiplicity and the critical accounts they offer of the other's conceptual presentation. I have insisted that a focused accounting of multiplicity as a *structure* clarifies both the affirmative positions and critical results developed in this conversation. Likewise, attention to the *procedure* prescribed by their respective structural commitments offers a more comprehensive accounting of the relation of one and multiple in their work, which situates their programs as alternatives to determining the nature of being; where it was formerly *One* or *many*, Deleuze and Badiou offer a novel option in their respective versions of *multiplicity*.

In Deleuze's work, the structure afforded by continuous manifolds makes way for the connection or expansion to *n*-dimensions in any local region of manifoldness; in this case, the dimensions are connections forged between elements of the virtual, and in this local region the demands for assessment and evaluation are prescribed by the connections themselves. In its affirmation of local evaluation, Deleuze's account of multiplicity develops apart from an ambient frame of reference. Approaching these commitments through their original articulation in work by Bernhard Riemann clarifies Deleuze's interpretation and deployment in his system. I attended to this structure in the definition of the virtual Idea in *Difference and Repetition*, demonstrating the Riemannian principles in its presentation, and in the relation of space

B. Vartabedian, *Multiplicity and Ontology in Deleuze and Badiou*,
https://doi.org/10.1007/978-3-319-76837-3_6

Deleuze and Guattari prescribe in "1440: The Smooth and the Striated" in *A Thousand Plateaus*.

With respect to procedures, both on his own and in collaboration with Guattari, I explained that the genetic structure of Deleuze's work is built using the differential elements *dx*, *dy*, their relation *dy/dx*, and the determination of this relation. I explained Deleuze's peculiar interpretation of these differential elements against the background of a more conventional assessment of the differential, which is used to identify average and instantaneous rates of change at any point on a curve. The so-called transcendental interpretation of the differential calculus brackets the existence of the differentials and in doing so, relieves the apparatus from having to locate these in space and time. I offered an interpretation of the social Idea in Deleuze's *Difference and Repetition*, which seats the conditions for any concrete transformation in trace elements of an idea circulating in any social context. I also discussed the presence of subtraction and interrogation deployed to "make the multiple" at the ground of any consolidated instance. The analysis, evident in the figures of the war machine as a-centered system and in an earlier accounting of overturned Platonism, reinforces the status of any consolidated instance as a kind of one-ness in suspension, always understood according to the multiplicity producing it. I concluded my analysis of structure and procedure in Deleuze and Guattari's work by evaluating "Example 1" in *What is Philosophy?* and its attendant diagram as an articulation of these concerns 'in action.'

In my discussion of Badiou's work, I showed that structure consists largely in the threefold relation of inconsistent multiple—void—consistent multiple. In Chapter 3, I discussed the way these are developed from developments in the discussions concerning set theory in the late nineteenth and early twentieth centuries, discussions involving Cantor, Frege, Russell, and Zermelo. I discussed the status of the inconsistent multiple as no-thing in Badiou's work, and the way this registers a break with Cantor's onto-theological assertion of an Absolute presence capable of closing inconsistency. I assessed Zermelo's axiom of separation, which prevents the overlap of consistent with inconsistent multiplicity, an occasion that would undermine the apparatus over which set theory operates. I identified under "procedure" those mechanisms that produce one-ness from multiple, including the account of the void, its mark $\emptyset$, and the axiom establishing it as first existent at the foundation of any ontological structure. I discussed the axioms of set theory used to construct these structures, and used this analysis of one-ing in *Being and Event* as a means to approach the presentation of scission in earlier work by Badiou; scission cuts any one from the multiple and follows from an understanding of the dialectic as division. I discussed Hegel's assessment of *something* appearing in *The Science of Logic* as a background for Badiou's interpretation, and I discussed the way this informs a procedure scission as one-ing, which he develops in *Theory of the Subject*.

In addition to their respective commitments to multiplicity and their deployments of mathematics, I argued that the attention, particularly to procedure, reveals a third continuity between Badiou's and Deleuze and Guattari's work in the form of subtractive analysis; this is analysis to identify the multiplicity, whether the set of precipitating connections (in Deleuze's case) or the founding void point (in Badiou's). To accomplish this, I developed some terminological precision around what I called a *general subtraction* in Badiou's work, distinguishing it with the work of following a truth through the pathways of special subtraction developed in the Gamma Diagram. From this common commitment, I put the resources developed under the auspices of structural and procedural analysis to the original terms of the conversation from 1976 to 1997; this revealed consistent critiques around the direction of multiplicity, the rigidity of a set-theoretic structure, and the very status of ontology itself as *productive* or *descriptive*. I concluded by initiating another, entirely indirect, conversation concerning the way mathematics functions in Badiou's and Deleuze's work, identifying the valences of rigidity and flexibility that attend their programs, respectively. The work of thinking through *model* as a methodological commitment forms a frontier for continued investigation.

To be sure, transformation is a front-line concern for both Badiou and Deleuze. As I discussed in the initial moves Chapter 2, much of what motivates Badiou's work is a search for traceable transformation, inspired by the failure of May '68 or the abdication of revolutionary politics toward more comfortable parliamentary procedures. When Deleuze and Guattari write "Rhizome," their concern is to overturn the commitments and tendencies present in traditional materialism; Deleuze, in *Difference and Repetition*, similarly assigns to philosophy the task of overturning Platonism (*DR* 59). I have acknowledged the presence of this expectation as a specter in my own work. However, instead of tracing their work in the direction of the new, I've seated their presentations in the direction of the old—that is, in the terms of an ancient ontological question concerning whether being is One or many.

Other Lineages

One of the benefits of reading the Badiou-Deleuze relation from the point of a common conceptual commitment like multiplicity is that it situates their apparent dispute in a longer lineage of philosophical investigation. The availability of *multiplicity* (precisely of the sort Badiou and Deleuze invoke) as an ontological choice draws these programs into conversation with Lucretius and the atomists, and Badiou's avowed mathematical Platonism offers a perennial disjunction against the overturned Platonism distinguishing Deleuze's account. Following the method I've suggested in the preceding pages, attention to the conceptual underpinnings of their respective deployments of multiplicity offers an additional dimension to these comparisons, which are themselves regular features of this particular philosophical landscape.[1] In all cases, the opportunity to liberate the Badiou-Deleuze knot from an internecine dispute and open it on to a broader landscape of philosophical inquiry adds depth to our understanding of—and appreciation for—these carefully-developed systems. Indeed, and as I have explained elsewhere, a recognition of the explicitly philosophical or ontological content of terminology emerging from the Badiou-Deleuze relation or in their respective programs adds value to their deployment in other fields (Vartabedian 2017).

Taking multiplicity as an ontological foundation in Badiou's work yielded a return to Kant's notion of the manifold (*Mannigfaltigkeit*). In "Kant's Subtractive Ontology," Badiou assesses the ontological prospects apparent in Kant's work, particularly in the assertion of the transcendental unity of apperception and the transcendental object = *X*. Badiou reads these as analogous to his own deployment of the void, holding a place at the foundation of cognition (in Kant's case) and ontology proper (in Badiou's). Though Badiou uses this occasion to argue his ontology is genuinely subtractive, insofar as it resolves to a *single* void and not the relation required by the "two voids" present in Kant's system, this complicates Badiou's assertions elsewhere that his ontology is an alternative to ontologies of presence. Because the void essentially overwrites inconsistent multiplicity, an act of faith in this infinite and uncountable no-thing must be invoked to maintain its presence beyond the boundary to though posited in the void. By re-seating Badiou's work in conversation with that of Kant, and by mobilizing the key signposts of his ontology and the content with which these are invested, I suggested that Badiou opens his theory to compromise according to principles he is keen to avoid. Put another way, in the move to compare his ontology with the programs of other philosophers (particularly those with whom more concrete continuities, such as multiplicity or subtraction, are absent), Badiou identifies a liability attending the severity with which he's restricted ontological inquiry from philosophical investigation.

This liability and limitation were not only developed in relation to Kant's work, however; as I showed in Chapter 5, it also shows itself in Badiou's relationship to Heidegger. The restriction Badiou enacts in *Being and Event* situates the mathematized ontology and its object in inconsistent multiplicity as an alternative to the abdication evident in Heidegger's "poetic overturning" (*BE* 27), which expresses the exhaustion of the more general onto-theological position; if the language and tools to express being qua being are *at the start* inadequate to their task, then it ought not be surprising that these find a limit. Philosophy and ontology are thus gutted, identified as (prematurely) at their end and ready to be abandoned. Badiou restricts ontological discourse to mathematics in *Being and Event*; however, when he approaches Heidegger at a meta-ontological level, without the rich resources of his mathematical ontology, liabilities and limitations in Badiou's project become apparent. This indicates, not necessarily the failure of Badiou's project, but the difficulty with which thinkers after Heidegger face when disentangling their positions from his.

For example, Knox Peden (2014) has pointed out that Deleuze's break from Heidegger is not decisive, and the reinvigoration of Spinozism as an alternative to phenomenology requires a Heideggerian presence to build against. Peden (2014, 217) explains, "In order to produce a Spinozism that is truly 'post-Heideggerian,' Deleuze must recuperate a notion of the infinite as anterior to the finite temporality regnant in Heidegger's exquisite account of modal existence." Put another way, it is not merely a bracketing of the subjective center that occupies phenomenology, but Deleuze's work requires another *inversion*. In the preceding chapter, I discussed the way in which a new accounting of materialism on the order of the rhizome requires a reversal of origins—the material from which any single instance emerges is connected, multiple, and ruptures in unpredictable though creative ways. In Peden's pithy description here, the reader finds an alignment of the 'anterior infinite' with the virtual, the depths to which interrogation, division, and subtraction dive to account for the conditions producing any finite instance.[2] Indeed, Peden characterizes the terms I have been interpreting according to a Riemannian framework (with a heritage in the Kantian notion of the manifold) as originating in Deleuze's disposition toward Spinozism. He identifies conceptual affinities between Deleuze and Althusser concerning a Spinozist interpretation of multiplicity (2014, 238), and approaches the structure of the Deleuzian idea and its differential elements from this vantage. Peden's approach to reading Deleuze, particularly in relation to Heidegger, to Althusser, and ultimately to Spinoza, opens yet another frontier for engaging the Deleuze-Badiou relation, specifically concerning multiplicity. Peden's insights identify a set of fissures and fractures that have deeper origins than the boundary I've identified in Kant's work.

New Frontiers

Readers attentive to the exchanges between Badiou and Deleuze in the period to which I've limited my inquiry may notice a curious absence in the preceding: Badiou's review of Deleuze's *The Fold: Leibniz and the Baroque*. In this review, Badiou's entire first section is devoted to an analysis of Deleuze's sense of multiplicity, and it prefigures much of what appears in *Deleuze: The Clamor of Being*. I have left this for last because in the brief opposition Badiou raises to Deleuze's theory of the multiple here is revealed the most fundamental determination to which each program is committed and a further frontier for exploration.

Badiou (1994, 52) points out, first, that the notion of the fold presents an "antiextensional concept of the multiple, a representation of the multiple as labyrinthine complexity, directly qualitative and irreducible to any elementary composition whatever." By this, Badiou seems to mean that Deleuze's multiplicity resists reduction to formal language and cannot be trapped under a logical concept; it manages some form of qualitative expression from a complex tangle of relations. This description also reveals (by implication) Badiou's own demands for a concept of the multiple: it *is* formalizable in logical language and has tools—an axiom of extensionality, for example—that indicate its in/equality with other multiples; it eschews connection and related complexity except for the most basic forms of relation, belonging and inclusion. When reinvention comes, it will be legible in an elementary composition, and as such reveal a fragment of being qua being that has been there all along. Perhaps only in amorous love will this event be expressed in a radically local way, though the steady state of Badiou's ontology will insist that this love is structured like all other great loves.

In my remarks concluding Chapter 5, I emphasized the 'virtue' of locality in Deleuzian multiplicity, a virtue that is used as one in a series of reasons for resisting the claim that Deleuze is a philosopher of the One-All. In his review of *The Fold*, Badiou (1994, 54) discusses the liability attending locality, saying:

> Nuance will be used to dissolve the latent opposition, one of whose terms the clear magnifies. Continuity can then be established locally as an exchange of values at each real point, so that the couple clear/obscure can no longer be separated, and even less brought under a hierarchical scheme, except at the price of a global abstraction. This abstraction itself is foreign to the life of the world.

Badiou's objection here is with Deleuze's apparent refusal to *choose among* or *select against*; where "the clear," those ideas following Plato and Descartes, relieves obscurity and offers a path to decision, Deleuze's interpretation of Leibniz in *The Fold*—and indeed much of what I've discussed in the preceding pages—rejects these in favor of any point's fragmentation into a multiplicity. Perhaps the abstraction Badiou is calling on here is something like the "inexact yet rigorous" notions Deleuze deploys in his work, but Badiou seems to suggest that even the most palatable abstraction would fail to generate sensible results. Deleuze is the philosopher of the fracture, of fragmentation, of the crack; these are conditions any count, any clear and distinct idea, or any Form would fail to collect to itself.

In a pithy statement of the stakes, at least as far as Badiou imagines these, he says, "Animal or number: this is the cross of metaphysics" (1994, 55). Here again, Badiou repeats the assignment of vital and mathematical multiplicities, an assignment that, as the review of Daniel W. Smith's argument in Chapter 2 proposed, Badiou uses to cover over the diverse mathematical commitments on which Deleuze's project depends. In Chapter 5, my attention to Badiou's work with Riemann in "One, Multiple, Multiplicities" suggests that because Deleuze does not follow Riemann *to the letter*—that is, to the crossroads of set theory and its ontological potential—Deleuze's multiplicity will only ever be vitalist. However, as an alternative to the cross of animal and number, I propose that an examination of multiplicity reveals that the cross of metaphysics—the lowest layer and fundamental determination—for Badiou and Deleuze is either *dissolution* or *decision*. This amounts to a revision of the terms of the debate in a novel framework, as Badiou's decision that "the One *is not*" accomplishes, or an inversion of the debate itself, as Deleuze's turn toward multiplicity as substantive requires. That is, we find ourselves at the frontier of determining how any ontology that has displaced subjectivity in the ways Deleuze and Badiou have gets underway in the first place.[3]

Readers of Badiou's 1988 review must fast-forward to 2011, when the Nirenbergs launch a similar-sounding critique at Badiou's ontology, claiming it empties ontology of being, of thought, or any other human characteristics that would make it "make sense" in the life of the world (after a phrase). But it seems fair to pose the same question to Badiou; not, as the Nirenbergs demand, to make ontology an endeavor seated with human subjectivity again, but rather to push its boundaries as far as possible. Isn't all of what Badiou and Deleuze have taken up an abstraction "foreign to the life of the world," and *isn't that the point*? The price of ontology conducted absent any subject raises the question, in the persisting attention to mathematical procedure, to structure, and to being and thought ambitiously loosed from the controls of subjectivity, Badiou and Deleuze each ask us to consider again *what do we mean when we talk about 'being'*? What abstraction, whether qua being or qua thought, will allow for imagining the new, the novel, or the transformed? In the preceding I have emphasized, not the event, but rather their respective understanding of multiple and its relationship to one as necessary for recognizing this possibility.

Notes

1. Take, for example, A.J. Bartlett (2011), Ryan J. Johnson (2016), John Bova and Paul Livingston (2017); Thomas Nail's recent return to Lucretius as "the first philosopher of immanence" (2017) brings Lucretius into another register with Badiou and Deleuze, theorists of 'immanent multiplicity.'
2. In *Pocket Pantheon*, Badiou (2009, 117–118) cites this commitment to infinity as precisely that which unites his work with Deleuze's.

In characteristic fashion, of course, Badiou insists it is a choice *against finitude* and its apparatuses, rather than an affirmation of infinity or, as Peden explains here, a matter of prioritizing infinity as Deleuze seems to do.

3. As Jay Lampert (2015, 272) has pointed out, even a decision for Deleuze is an instance of fracture: "When Deleuze speaks of the 'power of decision,' he means something like commitment to a serial problem. A decision is not 'decisive' in the sense that an argument or a battle is. It is not a state of mind, a set of reasons, or an outcome, but a divergence-point and all its possible branches, both fulfilled and unfulfilled."

References

Badiou, Alain. 1994. "Gilles Deleuze, *The Fold: Leibniz and the Baroque*." In *Gilles Deleuze and the Theater of Philosophy*, edited by Constantin V. Boundas and Dorothea Olkowski, 51–69. London: Routledge.

———. 2009. *Pocket Pantheon: Figures of Postwar Philosophy*. Translated by David Macey. London: Verso. Originally *Petit Panthéon Portatif*. Paris: Éditions la Fabrique, 2008.

Bartlett, A.J. 2011. *Badiou and Plato: An Education By Truths*. Edinburgh: Edinburgh University Press.

Bova, John, and Paul Livingston. 2017. "Univocity, Duality, and Ideal Genesis: Deleuze and Plato." In *Contemporary Encounters with Ancient Metaphysics*, edited by Abraham Jacob Greenstine and Ryan J. Johnson. Edinburgh: Edinburgh University Press.

Johnson, Ryan J. 2016. *The Deleuze-Lucretius Encounter*. Plateaus—New Directions in Deleuze Studies Series. Edinburgh: Edinburgh University Press.

Lampert, Jay. 2015. "Deleuze's 'Power of Decision', Kant's = X, and Husserl's Noema." In *At The Edges of Thought: Deleuze and Post-Kantian Philosophy*, edited by Craig Lundy and Daniela Voss, 272–292. Edinburgh: Edinburgh University Press.

Nail, Thomas. 2017. "Lucretius and the Immanence of Motion." *The Immanent Frame: Secularism, Religion, and the Public Sphere*. Social Science Research Council (SSRC). Available Online: https://tif.ssrc.org/2017/09/19/lucretius-and-the-immanence-of-motion/. Accessed 5 January 2018.

Peden, Knox. 2014. *Spinoza Contra Phenomenology: French Rationalism from Cavaillès to Deleuze*. Cultural Memory in the Present Series, edited by Hent de Vries. Stanford, CA: Stanford University Press.

Vartabedian, Becky with Nathan Eckstrand. 2017. "Early Career Research Spotlight: Becky Vartabedian." *Blog of the American Philosophical Association*. https://blog.apaonline.org/2017/11/15/early-career-research-spotlight-rebecca-vartebedian/. Accessed 15 November 2017.

Index

B. Vartabedian, *Multiplicity and Ontology in Deleuze and Badiou*,
https://doi.org/10.1007/978-3-319-76837-3

W

Z

The manufacturer's authorised representative in the EU is Springer
Nature Customer Service Centre GmbH, Europaplatz 3, 69115 Heidelberg,
Germany. If you have any concerns regarding our products, please
contact ProductSafety@springernature.com

Printed and bound by CPI Group (UK) Ltd, Croydon, CR0 4YY
05/10/2026
02228140-0001